Gas-Liquid and Liquid-Liquid Separators

Gas-Liquid and Liquid-Liquid Separators

Editor

A. K. Nigam

Gas-Liquid and Liquid-Liquid Separators

Edited by **A. K. Nigam**

Printed in 2017

ISBN: 978-1-68117-387-0

Library of Congress Control Number: 2015941575

© 2016 by

SCITUS Academics LLC,
616, Corporate Way, Suite 2, 4766,
Valley Cottage, NY 10989

www.scitusacademics.com

Contents

Preface

This practical guide is designed to help engineers and operators develop a "feel" for selection, specification, operating parameters, and trouble-shooting separators; form an understanding of the uncertainties and assumptions inherent in operating the equipment. The goal is to help familiarize operators with the knowledge and tools required to understand design flaws and solve everyday operational problems for types of separators. The most important gas/liquid separations that take place in oil field operation have been investigated. An inventory has been made of the conditions under which the separations have to take place and which requirements have to be fulfilled. The presently available separator types have been evaluated with respect to the suitability to fulfil the requirements listed above. It appeared that many separator types were not specifically designed for high pressure gas/liquid separation (rather for either atmospheric gas/liquid or high pressure gas/dust separation). It also appeared that in many cases the behaviour of the separator could not be reliably predicted under the conditions of the practical application.

Editor

Optimal Design of Drainage Channel Geometry Parameters in Vane Demister Liquid–Gas Separators

Fatemeh Kavousi[a], Yaghoub Behjat[b], and Shahrokh Shahhosseini[a]

[a]School of Chemical Engineering, Iran University of Science and Technology, Tehran, Iran
[b]Process Development and Equipment Technology Division, Research Institute of Petroleum Industry, Tehran, Iran

ABSTRACT

Vane liquid–gas demisters are widely used as one of the most efficient separators. To achieve higher liquid disposal and to avoid flooding, vanes are enhanced with drainage channels. In this research, the effects of drainage channel geometry parameters on the droplet removal efficiency have been investigated applying CFD techniques. The observed parameters are channel angle, channel height and channel length. The gas phase flow field was determined by the Eulerian method and the droplet flow field and trajectories were computed applying the Lagrangian method. The turbulent dispersion of the droplets was modeled using the discrete random walk (DRW) approach. The CFD simulation results indicate that by applying DRW model, the droplet separation efficiency predictions for small droplets are closer to the corresponding experimental data. The CFD simulation results showed that in the vane, enhanced with drainage channels, fewer low velocity sectors were observed in the gas flow field due to more turbulence. Consequently, the droplets had a higher chance of hitting the vane walls leading to higher separation efficiency. On the other hand, the parameters affect the liquid droplet trajectory leading to the changes in separation efficiency and hydrodynamic characteristic of the vane. To attain the overall optimum geometry of the drainage channel, all three geometry parameters were simultaneously studied employing 27 CFD simulation cases. To interpolate the overall optimal geometry a surface methodology method was used to fit the achieved CFD simulation data and finally a polynomial equation was proposed.

INTRODUCTION

Wave plate mist eliminators (vanes) are widely used in chemical, oil and gas industries to remove fine liquid droplets from gas flow. Fine liquid droplet separation is necessary for a number of reasons, such as reduction of pollutant emission into the environment, preventing damages to downstream equipments caused by corrosive or scaling liquid, recovering valuable products dispersed in a gas stream, increasing purity of gases for successive treatments and enhancement of the global operation economy.

Vane demisters can effectively remove entrained liquid droplets from a gas flow, usually by inertial impingement. The vane equipment basically consists of a number of narrowly spaced bended plates oriented in the direction of the gas flow as shown in Fig. 1. The droplet laden gas stream flows through the tortuous channels containing sharp bends and the flow direction changes repeatedly. The entrained droplets that are not able to follow these changes of direction, due to their inertia, deviate from the main gas flow and impact on the channel walls, where they coalesce and form liquid films that are continuously drained out from the separator by gravity.

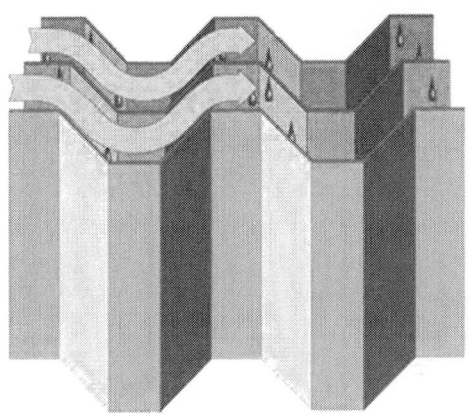

Figure 1: A horizontal vane demister.

The separation efficiency of a vane is affected by different factors, such as fluid inlet velocity, vane geometry and droplet mass fraction. Flooding due to separated liquid film accumulation on the walls and re-entrainment are also of concern. To overcome this issue and to increase droplet removal, the vanes are usually enhanced with drainage channels. The presence of drainage channels leads to an increase in separation efficiency and capacity of the system (Houghton and Radford, 1939, McNulty et al., 1987 and James et al., 2003).

The increasing need for cost and pollution control has generated great interest in vane demister performance studies. Early experimental and theoretical studies including those of Burkholz and Muschelknautz (1972) form the basis of the efficiency model for vane demisters currently used in the industry. In addition, Worrlein (1975) investigated separation calculations leading to the proposal that the optimal distance between the bends of a mist eliminator should be equal to the channel width. Ushiki et al. (1982) examined the performance of a separator with multistage rows of flat blades as described by Phillips and Deakin (1990), who have reported their experimental results on two vane demisters.

Wang and James (1998) simulated a wave-plate mist eliminator (without drainage channel), through the software CFX. The performance of low Re and STD k–ε models were studied and it was observed that the predictions were better applying the low Re turbulence model. In this survey, the comparisons between numerical predictions and experimental data showed a considerable difference due to neglected turbulent dispersion effects. In a later study (Wang and James, 1999) the effect of turbulent dispersion on droplet deposition was accounted for, using eddy interaction models (EIM). Modifications were made to the original model and it was proved that besides small droplets, the modified EIM resulted in more agreeable predictions when compared with the available experimental data.

Josang (2002) used various turbulence models when simulating the separating ability of the vane of their study. Having stated the importance of a co-ordinance between near wall treatment and the grids in their previous study, they compared predictions between structured and unstructured grid simulations. They have not reported any comparison with experimental data.

In the study of Jia et al. (2007) separation efficiency of a simple vane was compared with that of a vane with drainage channels. Among several re-entrainment mechanisms discussed, secondary droplet formation due to wall impingement of the droplets was taken into account. It was proven that drainage channels improve separation efficiency due to more turbulence and lower particle response time. Having studied a range of droplet diameters, it was concluded that generally for bigger droplets better separation was achieved, until re-entrainment was considerably increased.

The relation between separation efficiency and structural parameters of a demister vane was studied byZhoa et al. (2007) using the response surface method. The authors investigated the separation of droplets in the size range of 10–40 μm, considering only the drag force in the equation of motion, the effect of turbulent dispersion was neglected.

Galletti et al. (2008) studied the importance of eddy interaction models by simulating two vane demisters. The results of their study indicated that the EIM model was unable to predict realistic results for a range of droplet sizes. Comparisons of the results with some experimental data are reported in their paper. They suggested a modified EIM model as well as a model for turbulent flow in low velocities.

Recently, Narimani and Shahhosseini (2011) stated the importance of vane geometry on removal efficiency and pressure drop using CFD. In their study the effect of increasing gas velocity on vane demister separation efficiency has been analyzed. Among re-entrainment mechanisms, it was supposed that breakup of the droplets by their impingement on a liquid film and re-entrainment from the liquid film is likely to occur. They optimized vane geometries using response surface methodology. Their model was validated using published experimental data.

Having evaluated the results of the literature, the significant effect of turbulence eddies on the removal efficiency was recognized. Therefore, the urge of considering turbulent dispersion on droplet behavior in the gas flow was obvious. On the other hand, since effective drainage channel geometry design and its effect on separation efficiency have not been reported, in this research, investigation of turbulent dispersion effect on the liquid droplet behavior as well as optimal design of drainage channel geometry parameters; channel angle, channel height and channel length have been accomplished.

MODEL DESCRIPTION AND GOVERNING EQUATIONS

An Eulerian–Lagrangian approach was used to simulate the two-phase-flow gas-droplet system in the vane. The motion of liquid droplets was simulated by solving equations of motion of individual dispersed phase. The continuity, momentum and energy equations for the gas flow are as (Ranade, 2002):

$$\frac{\partial \rho_g}{\partial t} + \nabla \cdot (\rho_g v_g) = 0 \tag{1}$$

$$\frac{\partial}{\partial t} \rho_g v_g + \nabla \cdot (\rho_g v_g v_g) = \rho_g g + \nabla \cdot \vec{\tau}_g \tag{2}$$

$$\rho_g C_{Pg} \left(\frac{\partial}{\partial t} T_g + v_g \cdot \nabla T_g \right) = -\nabla \cdot (k_g T_g) q_g - H_{gl} \tag{3}$$

$\vec{\tau}_g$ is the stress tensor term and is defined as below:

$$\vec{\tau}_g = -p \delta_{ij} + \mu \left[\left(\frac{\partial u_i}{\partial x_j} + \frac{\partial u_j}{\partial x_i} \right) - \delta_{ij} \frac{2}{3} \frac{\partial u_k}{\partial x_k} \right] \quad (i, j, k = 1, 2, 3) \tag{4}$$

In order to close the set of equations, it is necessary to obtain the values of k (turbulent kinetic energy) and (turbulent dissipation rate). Local values of k and can be obtained by solving their transport equations. Turbulent kinetic energy is generated by extracting energy from the mean flow, and modeled using the assumption of turbulent viscosity. The generation term in the transport equation for represents vortex stretching by mean flow and fluctuating flow. The dissipation term in the transport equation for k is simply equal to. Robustness, economy, and reasonable accuracy of standard k– model for a wide range of turbulent flows explain its popularity in the industrial flow. The modeled form of transport equations for STD k– model can be written as:

$$\frac{\partial}{\partial t} (\rho_g k_g) + \nabla \cdot (\rho_g v_g k_g) = \nabla \cdot \left(\frac{\mu_{t.g}}{\eta_k} \nabla k_g \right) + G_k - \rho_g \varepsilon_g \tag{6}$$

$$\frac{\partial}{\partial t}(\rho_g \varepsilon_g) + \nabla \cdot (\rho_g v_g \varepsilon_g) = \nabla \cdot \left(\frac{\mu_{t,g}}{\eta_\varepsilon} \nabla \varepsilon_g\right) + \frac{\varepsilon_g}{k_g}(C_{1\varepsilon} G_k - C_{2\varepsilon} \rho_g \varepsilon_g)$$

(7)

The Lagrangian approach is used to determine droplet flow field by tracking the trajectories of the droplets. The initial distributions of droplet size and velocity are determined as the initial values. The water liquid droplet trajectories were computed individually, integrating the force balance on the liquid droplet, at the specified intervals during the fluid phase calculation. When the droplet moves in the gas flow, the forces acting on it include gravitational, buoyancy and drag force. This force balance equates the particle inertia with the forces acting on the liquid droplet that can be written as below.

$$\frac{dv_P}{dt} = F_D(v_g - v_d) + g(\rho_d - \rho)/\rho_d + F$$

(7)

$$v_g = v'_g + \ddot{v}_g$$

(8)

where F is an additional acceleration term such as virtual mass force, FD(vg − vd) is the drag force per unit particle. In the present work, the correlation proposed by Morsi and Alexander (1972) was used for estimating drag coefficient of the droplets. Morsi and Alexander (1972) provided a set of correlations for the drag coefficient as a function of particle Reynolds number, which responds fairly well in particle Reynolds range 0–5 × 10⁴. The coefficients where chosen to minimize the difference between the experimental data and analytical curves.

$$F_D = \frac{18\mu}{\rho_d D_d^2} \frac{C_D \, Re}{24}$$

(9)

$$C_D = a_1 + \frac{a_2}{Re} + \frac{a_3}{Re^2}$$

(10)

Once the primary gas flow field was calculated, the droplet motion, trajectory and deposition on the vane wall are computed. The effects of the droplets on the gaseous phase (i.e. momentum transfer and turbulence modulation) were taken into account unlike other studies. Such two-way coupling approaches are justified by the dilute characteristic of the flow leading to more accurate results.

The wall-film model allows a single component liquid drop to impinge upon a boundary surface and form a thin film on the wall surface. The model can be broken down into four major sub-models covering interactions during the initial impact with a wall boundary, subsequent tracking on the surfaces, calculation of the film variables, and coupling to the gas phase. In this study, the wall interaction is based on the work of O'Rourke and Amsden (2000), where four regimes of stick, rebound, spread and splash are considered based on the impact energy and the wall temperature. Below the boiling temperature of the liquid, the impinging droplet can either stick or spread or splash, while above the boiling temperature, the droplet can either rebound or splash. The criteria by which the regimes are partitioned are based on the impact energy and the boiling temperature of the liquid.

Turbulent Dispersion

The dispersion of the droplets due to turbulence in the gas phase can be predicted using the discrete random walk (DRW) model. The random walk model includes the effect of instantaneous turbulent velocity fluctuations on the droplet trajectories applying stochastic methods. In a turbulent gas flow, as indicated in Eq. (7), the droplet trajectory calculations are based on the mean fluid velocity. To account for the droplet turbulent dispersion, the instantaneous fluctuating gas flow velocity can be added as in Eq. (8).

In the DRW model, the random values of the velocity fluctuations are constant in a time interval calculated by eddy lifetime. Each eddy is characterized by a Gaussian distributed random velocity fluctuation, u, v, and w and time scale, e. The velocity fluctuations are defined by a Gaussian probability distribution as:

$$u' = \xi \sqrt{\overline{u'^2}}$$

(11)

where ξ is a normally distributed random number. Two viewpoints are available for the DRW model; the constant eddy lifetime and the random eddy lifetime. In the constant approach, the eddy lifetime is defined as a constant:

$$\tau_e = 2T_L$$

(12)

where TL is the time scale and can be approximated as:

$$T_L = C_L \frac{k}{\varepsilon}$$

(13)

In the random eddy lifetime point of view, e is defined as a random variation around TL using a random number (r) between 0 and 1.

$$\tau_e = -T_L \log(r)$$

(14)

Random calculations of e lead to more realistic simulations. The only input required for the DRW model is the integral time-scale constant, CL. In addition, constant or random eddy lifetime approach has to be chosen.

Response Surface Method

In this research, the response surface method was applied to find the optimal conditions of a vane mist eliminator in terms of channel height (x_1), channel length (x_2) and channel angle (x_3) in order to maximize separation efficiency. The response surface method fits a polynomial, as given in Eq. (15), into the obtained CFD simulation data and then employs the polynomial to find the optimal conditions (Zhoa et al., 2007).

$$Y = \beta_0 + \sum_{i=1}^{k} \beta_{ii} x_i^2 + \sum_{i=1}^{k} \beta_i x_i + \sum_{i<j} \beta_{ij} x_i x_j + e(x_1, x_2, \ldots, x_k)$$

(15)

where Y is the response, k is a variable, e is the error and $\beta i, \beta_{ii}$ and β_{ij} are the unknown parameters in the second order polynomial model.

COMPUTATIONAL DOMAIN AND OPERATING CONDITIONS

Fig. 2 illustrates the geometrical domain of the vane in which α = 45, λ = 118.5 mm and s = 25 mm. As shown in this figure, drainage channels are embedded at the bends to improve the overall operation of the vane and achieve higher liquid separation rate (Fig. 3).

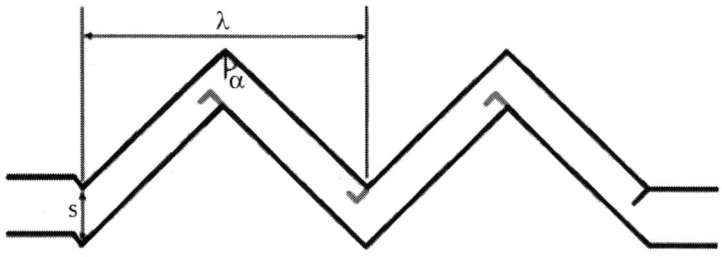

Figure 2: Vane demister geometry.

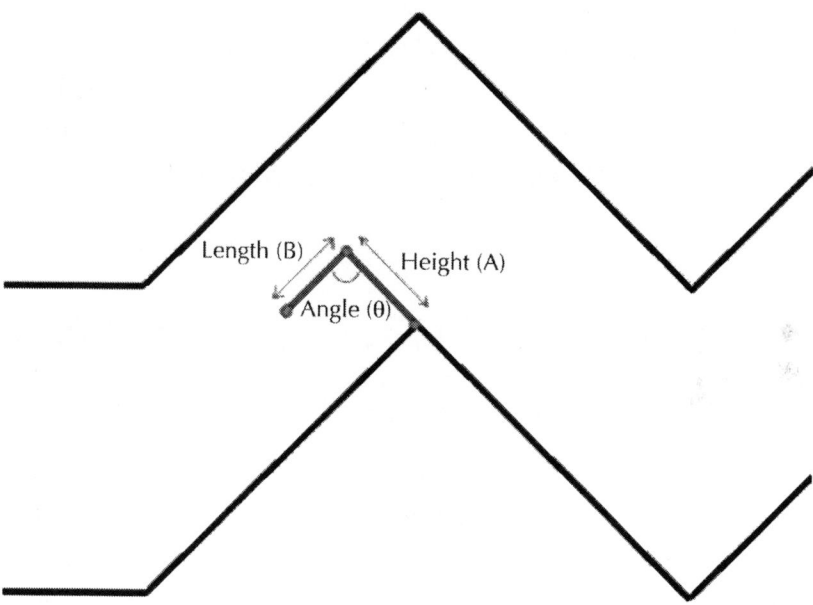

Figure 3: Drainage channel geometry parameters.

The three geometrical parameters investigated in this study are channel angle (θ), channel height (A) and channel length (B) as shown in Fig. 3. Initially, the height (A) of the channel and its length (B) were set to be 50% and 34% of the vane wall distance (12.5 mm and 8.5 mm), respectively. The channel angle (θ) was primarily 90°.

As an initial step toward using computational method to study fluid flow in the vane, the geometry was structurally meshed (Fig. 4). The calculation domain was divided into a finite number of control volumes (about 105,000 cells). Since the majority of separated droplets are expected to form a liquid film on the walls of the vane in comparison to the embedded channels, finer grids were generated near the wall region, to achieve precise results. In order to conserve computational time and still adequately provide accurate results, grid sensitivity studies were conducted to choose final grids that gave grid-independent numerical results upon further grid refinement.

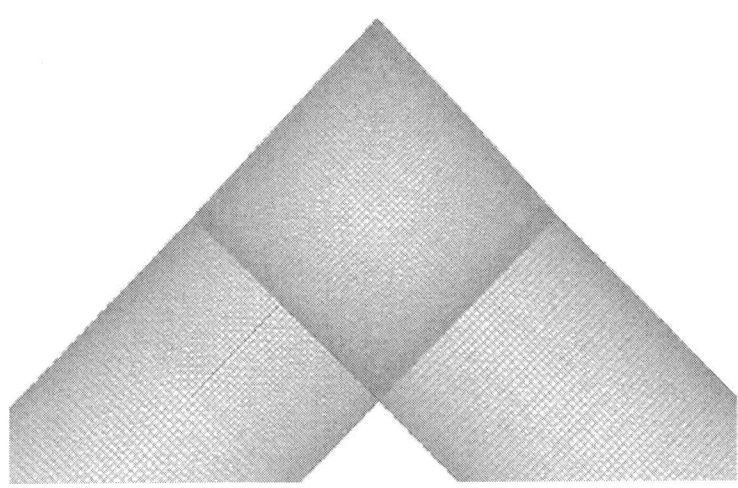

Figure 4: Vane computational grids.

The dispersed and continuous phases were water droplets and air respectively. Hydrodynamic and heat transfer conditions of the two phase flow (gas–liquid) are summarized in Table 1. The gas and droplet velocity in the vane inlet were equal (zero slip velocity with respect to the gaseous phase). The mass fraction of the droplets at the inlet was 10% corresponding to wetness of 0.09.

Table 1: Hydrodynamic and heat transfer conditions of the two phase flow

Flow pattern	P (MPa)	T (°C)	ρg (kg m^{-3})	ρd (kg m^{-3})	μg (μpas^{-1})	μd (μpas^{-1})
Dispersed flow	0.1	20	1.2	998	18	998

Simulation Assumptions

- Since in practice the depth of the vane is much larger than the other two dimensions, the flow is assumed to be two dimensional. Thus the two dimensional results could be extended to three dimensional calculations will negligible errors.

- Since the value of the Weber number in this study was below the critical value, film breakup was neglected.
- The droplet–film interaction at the walls is negligible.
- Once the droplets collide with the walls, they are immediately drained down, so they do not rebound into the gas flow.
- Re-entrainment was also not taken into account.
- The walls of the vane are assumed to be stationary and to have no slip shear conditions. The temperature of the walls is the same as the flow thus no heat transfer is happening.

It is assumed that the droplets with a uniform diameter of D_{di} were injected at the vane inlet and it is possible to find some droplets at the outlet with a diameter equal to or smaller than D_{di}. Thus, the separation efficiency of a droplet can be calculated as follows.

$$\eta = \frac{\sum_{i=1}^{n}(m_i \eta_{di})}{\sum_{i=1}^{n} m_i}$$

(16)

$$\eta_{di} = \frac{y_i}{x_i}$$

(17)

Governing equations of the developed model are solved by finite volume method employing Semi Implicit Method for Pressure Linked Equations (SIMPLE) algorithm (Patankar, 1980 and Versteeg and Malalasekera, 1995) that is developed for multiphase flow using Partial Elimination Algorithm (PEA). To achieve a high spatial accuracy the second order upwind scheme was used to discrete the equations. The conservation equations were integrated in space and time. The sets of algebraic equations were solved iteratively using an explicit method. The procedure was separated as two major iteration routines. The first one was for continuous phase (gas phase) flow calculation, and the second one was for the liquid droplet flow calculation (Patankar, 1980 and Versteeg and Malalasekera, 1995).

RESULTS AND DISCUSSION

In the first step of this research, the drainage channel effects on the overall hydrodynamic and performance of vane was investigated. The gas phase flow fields for a vane with drainage channels and one without the channels are shown in Fig. 5. The CFD simulation results in this figure display less low-velocity sectors in the vane with drainage channels due to more turbulence, created owing to these channels. The higher overall velocity in the channel embedded vane, meaning higher turbulence, is evident in this figure. As can be seen in Fig. 5, the gas phase is inclined to vane wall in the case of vane with drainage channels so droplet collision to wall is raised leading to higher separation efficiency.

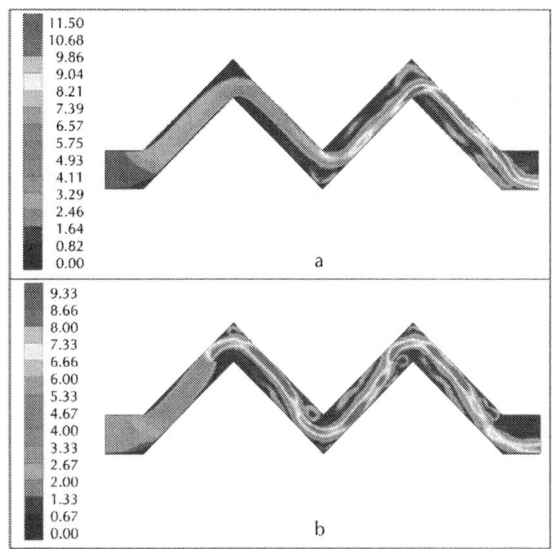

Figure 5: Gas phase velocity profile in the vane. (a) Vane without drainage channels. (b) Vane with drainage channels.

The effect of drainage channels to enhance the vane in terms of droplet trajectories is shown in Fig. 6. This figure illustrates that, in the case of vane with drainage channels, the droplets have a higher chance of hitting the vane walls leading to higher separation efficiency. The obtained result for droplet trajectories in Fig. 6 are accordance with the reported results for gas phase flow field that gas phase inclined to vane wall in the case of vane with drainage channels as shown in Fig. 5. This argument has been confirmed with the computation of the overall separation efficiencies for the two vanes with the same operating conditions (inlet velocity 2 m/s, 10% liquid droplets at inlet). The vane, enhanced with drainage channels, has an efficiency of 72.22% as compared to the 23.3% for the plain vane. On the other hand, pressure drops of the enhanced vane and the plain vane were 141.85 Pa and 27 Pa, respectively. Therefore, it is obvious that drainage channels have an impact on the vane performance.

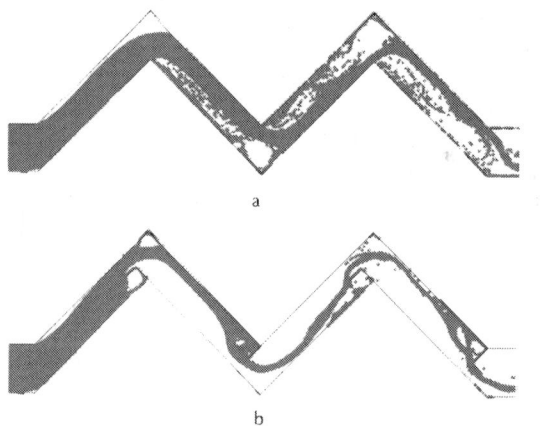

Figure 6: Droplet trajectories in the vane. (a) Vane without drainage channels. (b) Vane with drainage channels.

The developed computational model was applied to predict separation efficiencies of the vane using liquid droplets in the size range of 3–15 μm. To make comparisons between separation efficiencies for different droplet diameters, the droplets were assumed to be injected with a uniform size at the vane inlet in each simulation. The results were compared with the corresponding experimental data of Galletti et al. (2008) in order to validate the developed computational model as shown in Fig. 7.

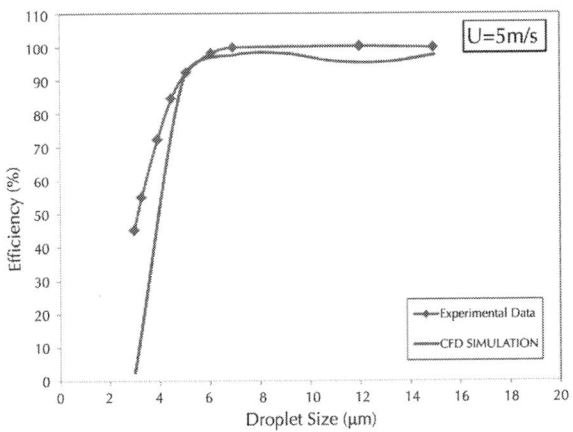

Figure 7: Comparison between predicted separation efficiencies and the experimental data.

At the vane inlet, the mass fraction of the droplets and velocity were 10% and 5 m/s, respectively. As can be seen in Fig. 7, the CFD model predictions and the experimental data acceptably agree. The results of this computational model show an increase in separation efficiency with the rise in droplet diameter. As it can be observed, for larger droplets (diameters more than 5 μm) predictions are more consistent with the experimental data.

To reduce the model error for fine droplet sizes (<5 μm) and to be more realistic in simulating this separation process, it is important to consider the effects of eddies on droplet deposition, using a turbulent dispersion model. Therefore, the discrete random walk model (DRW) was applied in the simulations of the small particles in order to evaluate the effects of adding turbulent dispersion.

Fig. 8 indicates that as a result of applying DRW model, the predictions for small droplets (<4.5 μm) are closer to the corresponding experimental data. Table 2 displays the relative errors of the simulation prediction for droplet separation efficiency applying DRW model and without considering the DRW model. As shown in this table the relative error for small droplets (<4.5 μm) is less applying DRW model. The reason is that smaller particles have less droplet momentums due to their lighter mass and are more likely to change with the velocity fluctuation term being considered in DRW model. The figure also demonstrates that for bigger droplets considering the effect of velocity tolerances in the DRW model deteriorates the model predictions. In addition, it is a fact that DRW model needs more calculation time. Therefore, it should be implemented for smaller particles.

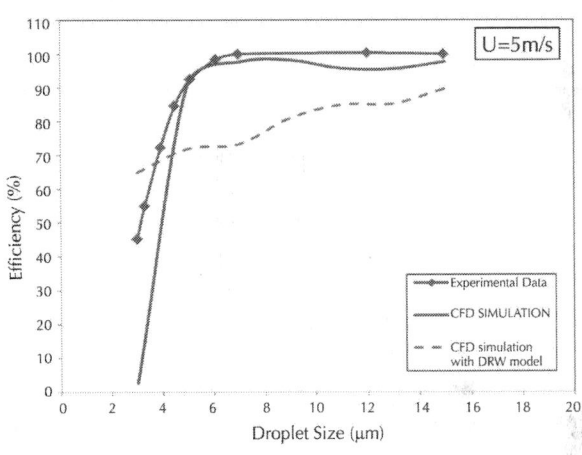

Figure 8: Comparison between predicted separation efficiencies and the experimental data.

Table 2: Relative error of the simulation results to predict droplet separation efficiency (%) with DRW model and without DRW

Droplet diameter (μm)	With DRW	Without DRW
3	85.714	93.333
3.2	22.641	77.358
3.5	8.064	55.0247
3.8	1.470	38.235
4	6.756	29.729
4.2	11.538	21.794
4.3	13.580	19.753
4.4	15.662	15.662

Different time scale constants were applied in the computational model applying DRW model. The computations were carried out for droplet sizes of 7 μm and inlet velocity of 2 m/s. Separation efficiency calculations are presented in Table 3. The results show that the separation efficiency does not change very much with moderate changes in time scale constant (CL).

Table 3: The effect of time scale constant on separation efficiency

CL	0.16	0.201	0.25	Constant eddy life time
Calculated separation efficiency	74.2	73.26	76.3	77.2

As discussed above, to improve the overall operation of the vane and achieve higher liquid separation, drainage channels were embedded at the bends. The three geometrical parameters investigated in this study are channel angle (θ), channel height (A) and channel length (B) as shown in Fig. 3. Primarily, each of the geometry parameters A, B and θ were studied separately as to evaluate the effects of changing their measures on the separation efficiency of the vane and the pressure drop of the flow. In other words, two of the three parameters were kept at their initial values and the third one was given different values each time.

Fig. 9 displays the effects of parameters A, B and θ on trajectories of the droplets. The results indicate that these parameters affect the liquid droplet trajectory leading to the changes in separation efficiency and hydrodynamic characteristic of the vane.

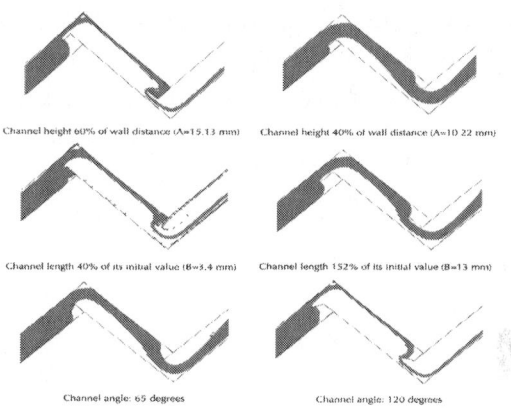

Figure 9: Droplet trajectories in the vane with two different channel geometry parameters.

In the first step, the channel height was changed to 40%, 45%, 50%, 55% and 60% of the wall distance, namely, 10.22, 11.22, 12.5, 13.96 and 15.13 mm, respectively. The droplet separation efficiencies were predicted using the developed CFD model, while keeping both the channel length and the angle constant at their initial values. Liquid droplet separation efficiency and pressure drops for different heights of the drainage channel are illustrated in Fig. 10.

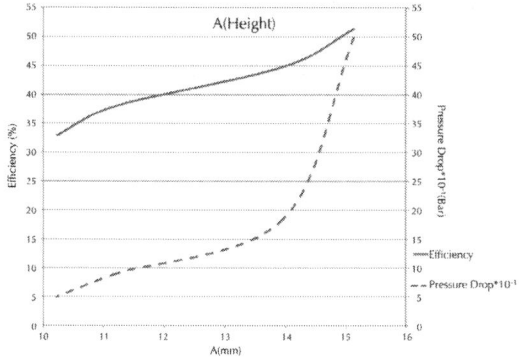

Figure 10: Pressure drop and separation efficiency versus drainage channel height.

This figure reveals that raising channel height causes separation efficiency to increase. The figure also shows a similar effect for pressure drop in lower ranges of the height. However, for A > 13 mm the pressure drop changes exponentially with the height.

In the circumstances where no special considerations regarding pressure drop of the flow or the desired separation efficiency exist, according to the graphs in Fig. 10, it can be presumed that the optimal height of the drainage channel is where the vertical distance between the pressure drop and efficiency graphs (Obj) is maximum, meaning that at a specific channel height, the efficiency is at its highest amount and at the same time the pressure drop is at its lowest:

$$Obj = E - \Delta P$$

(18)

where E and ΔP are the overall separation efficiency and pressure drop at a specific channel height respectively. Using this concept, at different channel heights, the vertical distance between the two graphs (the difference between pressure drop and separation efficiency (Obj)) was calculated. Referring to these plots, the optimal height for the drainage channel of this study is 12.2 mm, resulting in a droplet separation of 40.82% and a pressure drop of 111.5 Pa.

In the second stage the length of the channel (B) was studied. The length was changed from 40% to 152% of its initial value and the effects of this change on the efficiency and pressure drop were observed. The CFD simulation results are shown in Fig. 11.

According to the CFD results of Fig. 11, the efficiency curve has a maximum value of 80.8% when the channel length is 7.6 mm. The maximum takes place where the vane length is near to 90% of its initial values. However, due to the importance of pressure drop in most chemical processes, the optimal length of the channel is where efficiency is high and pressure drop is low, simultaneously. The vertical distance between the efficiency and pressure drop graph (Obj = E − P) was calculated in order to obtain optimum length of drainage channel. According to the calculated results, the overall optimum length of the channel is 8.5 mm, where 79.29%

of the laden droplets are separated and the fluid flow pressure drop is 118.1 Pa.

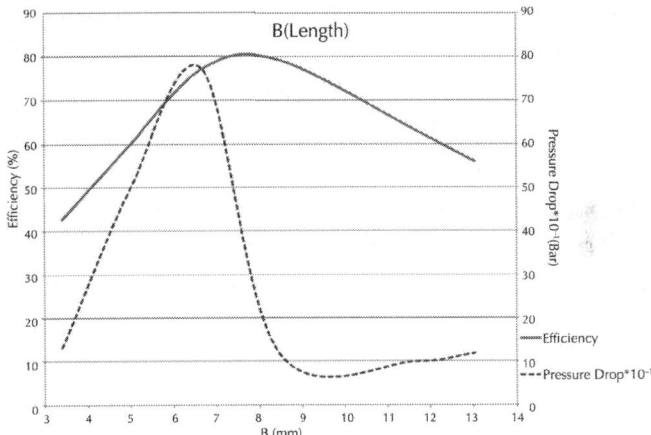

Figure 11: Separation efficiency in different channel lengths.

In the last stage, the effect of the channel angle was investigated using four different channel angles, 65°, 80°, 90° and 120°. The computed efficiencies and pressure drops are depicted in Fig. 12.

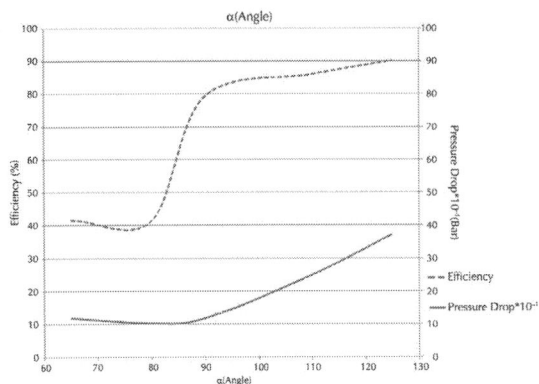

Figure 12: Separation efficiency and pressure drop applying different channel angles.

Fig. 12 shows the efficiency and pressure drop increase owing to the escalation of the channel angle. As in the cases of the other two geometry parameters, both separation efficiency and pressure drop were considered in order to determine the optimal channel angle. The figure implies that the optimal value of channel angle is around 95° where high percentage of separation (83.75%) and the low pressure drop (143.3 Pa) occur at the same time.

To attain the overall optimum geometry of the drainage channel, all three geometry parameters were considered simultaneously. Assuming 3 different values for each of the parameters (length, height and angle), 27 simulation cases were designed as shown in Table 4.

Table 4: Separation efficiencies and pressure drop predictions of 27 cases with different parameters

Case no.	A (mm)	B (mm)	α	Efficiency (%)	Pressure drop (Pa)
1	12.5	8.45	90	63.71	58.7
2	12.5	8.45	80	41.42	50.9
3	12.5	8.45	125	90.16	185.64
4	12.5	6.7	90	77.23	381.63
5	12.5	6.7	80	56.5	57.18
6	12.5	6.7	125	65.35	193.83
7	12.5	13.04	90	56.17	59.1
8	12.5	13.04	80	55.75	54.92
9	12.5	13.04	125	50.51	158.17
10	11.21	8.45	90	38.1	45.45
11	11.21	8.45	80	48.86	38.06
12	11.21	8.45	125	54.23	100.73
13	11.21	6.7	90	42.43	37
14	11.21	6.7	80	42.38	62.71
15	11.21	6.7	125	50.38	116.15

16	11.21	13.04	90	40.65	37.42
17	11.21	13.04	80	88.22	37.67
18	11.21	13.04	125	63.39	97.48
19	13.96	8.45	90	44.95	92.05
20	13.96	8.45	80	60.51	120.22
21	13.96	8.45	125	69.67	369.82
22	13.96	6.7	90	62.23	116.1
23	13.96	6.7	80	49.42	93.37
24	13.96	6.7	125	68.4	415.03
25	13.96	13.04	90	57.57	108.63
26	13.96	13.04	80	53.01	93.49
27	13.96	13.04	125	63.1	319.82
IOP[a]	12.5	8.45	95	84.3	54.9

[a]Case (IOP) is the geometry consisting of the individually optimized length, height and angle of the channel introduced in Fig. 10, Fig. 11 and Fig. 12.

The same operating conditions were applied for all the 27 simulations; inlet velocity of 2 m/s, inlet droplet mass concentration of 10%, droplet diameter of 7 μm, using DRW turbulent dispersion model with 0.201 time scale and random eddy lifetime. The STD k– model was used for the turbulent flow. The predicted efficiencies and pressure drops for the 27 cases are given in Table 4.

The results for a vane with individually optimized characteristics (channel height, channel length and channel angle respectively equal to 12.2 mm, 8.5 mm and 95°) were included in Table 4 as case (IOP). These results indicate that a vane with a geometry composed of individually optimized values for length, height, and angle does not necessarily give the best performance.

The CFD simulation results shown in Table 4 indicate that the highest achieved efficiency is 90.16% corresponding to the third case in the table, while the pressure drop is significantly high (185.64 Pa). Table 4also shows the lowest pressure drop (37 Pa) belongs to simulation 13, whereas the efficiency is 42.43%. Therefore, none of these cases can be the optimal design for drainage channels. Referring to the obtained results in Table 4, taking into account both separation efficiency and pressure drop of the vane, case 17 is the optimal design for the drainage channel in the vane, with the efficiency and pressure drop being 88.2% and 37.67 Pa. In this case the length and the height of the channel are 150% and 45% of their initial measures and the channel angle is 90% of its initial value.

It is clear that predicting the performance of all the possible geometries is quite costly and time consuming. To interpolate the overall optimal geometry for the drainage channels in the vane of a surface methodology was used to fit the obtained data, given in Table 3, to the below polynomial equation.

$$Y_{Obj} = a + bx_1 + cx_2 + dx_3 + ex_1^2 + fx_2^2 + gx_3^2$$

$$+ hx_1x_2 + jx_2x_3 + kx_1x_3$$

$$(19)$$

where x_1 is the channel height (A), x_2 is the channel length (B) and x_3 is the channel angle (ϑ). The coefficients (a) to (k) were determined as below.

a = −2.32877E+03	b = 1.83076E+02
c = 1.01692E+02	d = 2.09878E+01
e = −2.28068E+00	f = −4.55143E+00
g = −1.25497E−02	h = −1.96968E−01
j = −2.58979E−03	k = −1.70937E+00

The above regression makes it possible to study the three parameters simultaneously. In each of Fig. 13, Fig. 14 and Fig. 15, the objective function (Obj = E − ΔP) is plotted the geometry parameter.

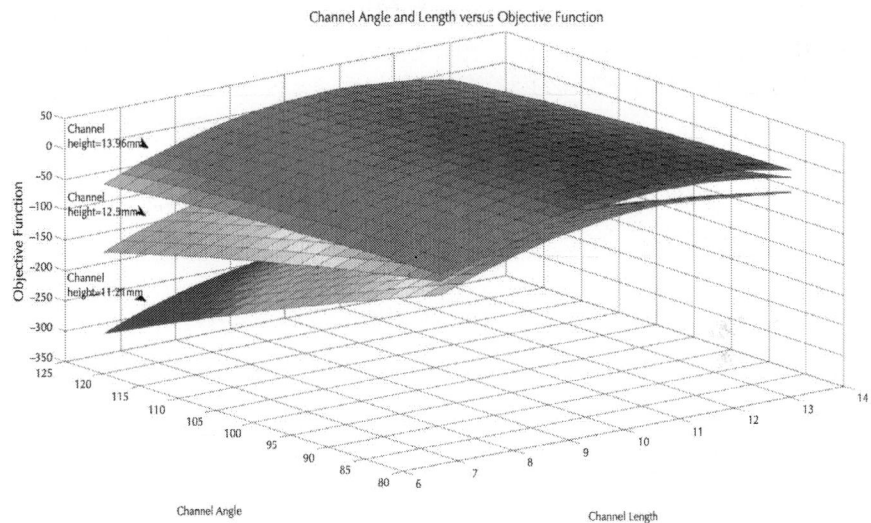

Figure 13: Objective function versus length and angle of the drainage channel.

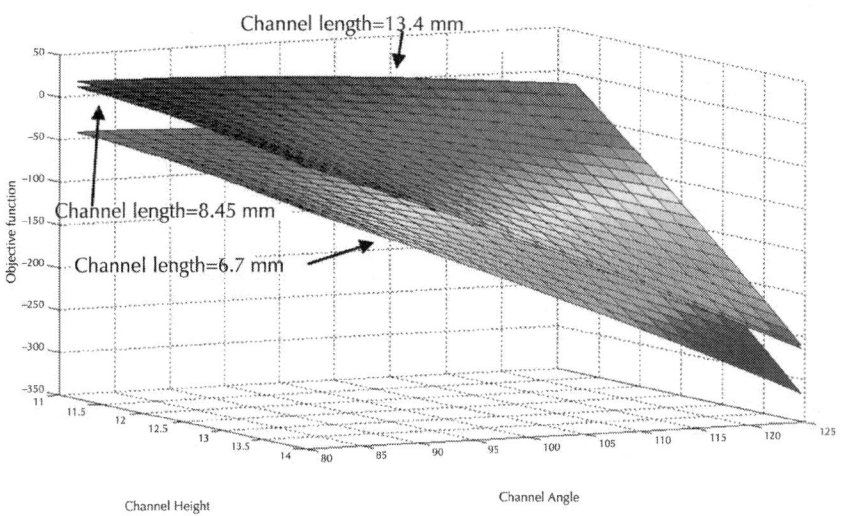

Figure 14: Objective function versus height and angle of the drainage channel.

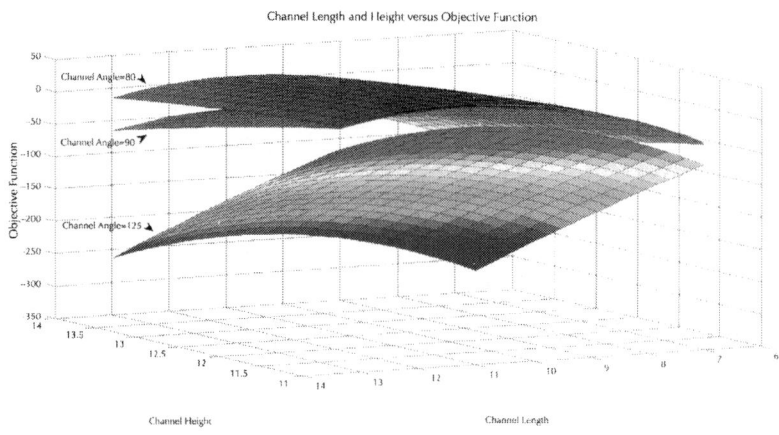

Figure 15: Objective function versus height and length of the drainage channel.

Fig. 13 illustrates the objective function increases with the rise of channel angle. However, the slope of this rise reduces for longer channels. This is while the changes in the objective function are not as big with the variations in the channel length for a constant channel height.

Fig. 14 elucidated that at certain channel lengths, the objective function declines when the vane angle exceeds a certain degree (~95°). The reason is the effect of channel geometry on the velocity profile of the flow and the channel ability to keep the collided droplets in the hook shaped channel, which results in efficiency drop. At channel angles near 95°, not only the channel is still in its hook shape to avoid the separated droplets to re-entrain into the gas flow and help them drain, but also due to its geometry, it considerably changes the turbulant regime of the fluid flow, causing more droplet impingment. It can also be concluded that the performance of the overall vane and channel system is improved for longer channels. Howevere, this increase is inconsiderable for longer channels, since very small performance improvement was achieved by changing the channel length from 8.45 mm to 13.4 mm. Fig. 15 shows for smaller channel heights, both the objective function and the separation efficiency increase when the channel length increases. However, this is not the case for bigger channel heights, in which, lower separation is predicted for bigger channel lengths. In the case of simultaneous very small length and height, the figure shows that the channel is not effective in changing droplet trajectories or capturing them. In addition, for large channels with long lengths and big heights low droplet separation is predicted. The predictions are justifiable since there might be a case of blocking the vane when the channel length and height exceed a certain value. This explains the importance of optimizing the channel design to achieve high separation efficiencies.

CONCLUSIONS

In this research, computational study has been performed to investigate the impacts of the drainage channel geometry parameters on droplet removal efficiency in wave plate mist eliminators. In the developed CFD model, the hybrid Eulerian–Lagrangian approach was used to simulate the gas-droplet two phase flow in the vane. The motions of liquid droplets were simulated by solving the equations of motion of the individual dispersed phase. Once the primary gas flow field was calculated, the droplet motion, trajectory and deposition on the vane wall are computed. The effects of the droplets on the gas phase were also taken into account. The dispersion of the droplets due to gas phase turbulence was predicted using the discrete random walk (DRW) approach. The CFD simulation results indicate that enhancing vane demisters with drainage channels leads to an increase in flow turbulence as well as reducing droplet re-entrainment. The increased turbulence obliges droplets to change trajectories and thus their impingement to the walls rises causing the overall droplet removal efficiency to increase. On the other hand, the design parameters of drainage channel affect the liquid droplet trajectory leading to the changes in separation efficiency and hydrodynamic characteristic of vane. The developed computational model was validated by comparing the CFD simulation predictions of separation efficiencies with the corresponding experimental data. The obtained simulation results indicate that applying DRW approach causes computational model predictions for small particles (<4.4 μm) to be closer to the corresponding experimental data. Whilst considering DRW approach in computational model for the bigger droplets leads to deterioration of the model perditions. The CFD simulation results revealed that raising channel height and channel angle caused separation efficiency and pressure drop to increase. Whilst the droplet removal efficiency and pressure drop escalated toward a maximum value by increasing channel length and then decreased. In order to achieve overall optimum geometry of the drainage channel, all of the geometry parameters were simultaneously considered by conducting 27 simulation cases. The optimum design for the drainage channel is defined as where the droplet separation efficiency is relatively high and the pressure drop is as low as possible. This optimal design was gained using the response surface method and fitting the appropriate equation into the results of the simulations.

REFERENCES

1. Burkholz, A., Muschelknautz, E., 1972. Tropfenabscheider. Chem. Ing. Tech. 44 (8), 503–509.

2. Galletti, Ch., Brunazzi, E., Tognotti, L., 2008. A numerical model for gas flow and droplet motion in wave-plate mist eliminators with drainage channels. Chem. Eng. Sci. 63, 5639–5652.

3. Houghton, H.G., Radford, W.H., 1939. Measurements on eliminators and the development of a new type for use at high gas velocities. Trans. Am. Chem. Eng. 35, 427–433.

4. James, P.W., Wang, Y.I., Azzopardi, B.J., Hughes, J.P., 2003. The role of drainage channels in the performance of wave-plate mist eliminators. Chem. Eng. Res. Des. 81, 639–648.

5. Jia, L., Suyi, H., Xiaoma, W., 2007. Numerical study of steam-water separators with wave type vanes. Chin. J. Chem. Eng. 15, 492–498.

6.   Josang, A., 2002. Numerical and experimental studies of droplet gas flow. Doctoral Dissertation in Chemical Engineering, Technology Telemark University, Norway.

7.   McNulty, K.J., Monat, J.P., Hansen, O.V., 1987. Performance of commercial chevron mist eliminators. Chem. Eng. Prog. 83, 48–55.

8.   Morsi, S.A., Alexander, A.J., 1972. An investigation of particle trajectories in two-phase systems. J. Fluid Mech. 55, 193–208.

9.   Narimani, E., Shahhosseini, S., 2011. Optimization of vane mist eliminators. Appl. Therm. Eng. 31, 188–193.

10.   O'Rourke, P.J., Amsden, A.A., 2000. A Spray/Wall Interaction Sub model for the KIVA-3 Wall Film Model, SAE Paper, 01-0271.

11.   Patankar, S.V., 1980. Numerical Heat Transfer and Fluid Flow. Hemisphere Publishing Corp., Washington, DC.

12.   Phillips, H., Deakin, A.W., 1990. Measurements of the collection efficiency of various demister devices. In: Proc 4th Annual Meeting of the Aerosol Society, Loughborough, UK.

13.   Ranade, V.V., 2002. Computational Flow Modeling for Chemical Reactor Engineering. Academic Press, London.

14.   Ushiki, K., Nishizawa, E., Beniko, H., Iinoya, K., 1982. Performance of a droplet separator with multistage rows of flat blades. J. Chem. Eng. Jpn. 15, 292–298.

15.   Versteeg, H.K., Malalasekera, W., 1995. An Introduction to Computational Fluid Dynamics: The Finite Volume Method, Second ed. Addison-Wesley Longman, Edinburgh, England.

16.   Wang, Y.I., James, P.W., 1998. The calculation of wave-plate demister efficiencies using numerical simulation of the flow field and droplet motion. Chem. Eng. Res. Des. 76, 980–985.

17.   Wang, Y.I., James, P.W., 1999. Assessment of an eddy-interaction model and its refinements using predictions of droplet deposition in a wave-plate demister. Chem. Eng. Res. Des. 77, 692–698.

18.   Worrlein, K., 1975. Drukverlust und fractionabshceidegrad in

periodisch gewinkelten kanalen ohne einbauten. Chem. Ing. Tech. 47.

19.   Zhoa, J., Jin, B., Zhong, Z., 2007. Study of the separation efficiency of a demister vane with response surface methodology. J. Hazard. Mater. 147, 363–369.

Chapter 2

Flow Modeling of a Battery of Industrial Crude Oil/Gas Separators Using113min Tracer Experiments

Hector Constant-Machado[a], Jean-Pierre Leclerc[b],
Eddie Avilan[a], Gustavo Landaeta[a],
Nelkys Anorga[c], Oscar Capote[d]

[a]Departamento de F´ısica Aplicada, Ingenier´ıa, Universidad Central de Venezuela, AP 47724, Caracas 1040, Venezuela

[b]Laboratory of Chemical Engineering Sciences, CNRS-ENSIC, 1rue Grandville, BP 451, 54001Nancy, France

[c]Universidad de Oriente, Venezuela d Petr ´oleos de Venezuela (PDVSA), Distrito Norte, Oriente, Venezuela

ABSTRACT

The objective of this work is to study the flow behavior of crude oil in a battery of industrial crude oil/gas separators in oil industry. The battery is composed of three separators operating at different pressures. The residence time distribution (RTD) of the crude oil has been determined by an impulse injection of[113m]In at the inlet of each separator and the concentration has been continuously recorded at the outlet. The real volume occupied by the crude oil has been determined by simple estimation of the first moments. The RTD of the crude oil has been simulated by a model composed of few mixing cells in series representing the effect of the deflector located at the entrance and a plug flow partly due to the high viscosity of the crude oil. The variation of the parameters of the model has been studied as a function of pressure conditions and they have been linked to the deposition of sediment. The tracer measurement in gas phase showed that this method can detect non-negligible malfunction but it is not sensitive enough to assume that the zero tolerance of oil concentration in gas phase requested by the exploitant is respected.

GENERAL DESCRIPTION OF CRUDE OIL/GAS SEPARATORS

The crude oil extracted from reservoirs contains also sediments, water and gas that must be removed before the beginning of the exploitation process. The large differences between physical properties of each component of the mixture facilitate the separation of sediments, water and gas from the crude oil. It is carried out in several separators at different temperatures of the pressurized fluids coming from reservoirs or homogenous tanks containing crude oil from different origins. This study concerns an exploitation located at the north of Monagas, Venezuela. The crude oil/gas separation is achieved in a battery of three separators. The separators work at different pressures to optimize the crude oil/gas separation so that as much as possible of the heavier hydrocarbon components are removed from the gas. Thus a greater quantity of the oil phase is obtained. Each gas line is connected to a tank that collects the outlet of many separations systems. Finally the gas is sent to the compressor plant of Amana in Monagas. After the different separators, the crude oil at atmospheric pressure is sent to Patio de Tanques Travieso in Monagas. The condensate fraction in the gas phase should be entirely eliminated whereas the water fraction in the oil phase has to be minimized under a limited value of 2%.

The battery of crude oil/gas separator is described Fig. 1. The crude oil is first partially separated from the gas in a first separator. At the outlet, it is mixed with oil coming from storage tanks. Then it goes through the second separator and the third one. The main operating parameters of the three separators are given inTable 1. The crude oil and gas flow rates have been predicted by a process simulation software package (PRO II).

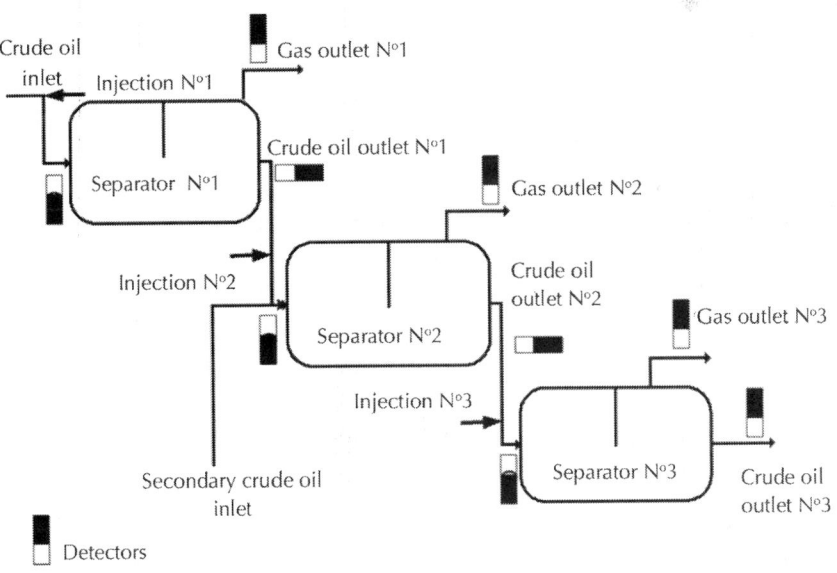

Figure 1: Schematic description of the battery of crude oil/gas separators.

Table 1: Main operating parameters in the different separators (separator 2 is connected with a secondary crude oil inlet which explains the higher flow rate)

Separator No.	Pressure (Pa)	Temperature (°C)	Crude oil flow rate, Q_{oil} (m³/s)	Gas flow rate (m³/s) NTP
1	8.37×10^6	60	0.0258	7.80
2	3.45×10^6	54	0.041	7.92
3	4.14×10^5	49	0.0375	4.13

The efficiency of the separators can be strongly affected by foam. Because of this, the exploitants need to add continuously chemical foam inhibitors. The quantity of inhibitors is adjusted by continuous control of the level inside the separator. Nevertheless, the foam remains present despite the inhibitors, and the flow rate of crude oil is adjusted by a control of the global level in the separator. The crude oil contains also some sediments that settle inside the separators. The separator contains four phases:

- A deposition of sediments (from 2 to 10 cm of thickness);
- The crude oil (from 10 to 20 cm of thickness);
- The foam (from 20 to 40 cm of thickness);
- The gas phase.

To get higher productivity, the thickness of the foam must be as low as possible. The level of crude oil inside the separator is an unknown important operating parameter. For a given flow rate, the volume of crude oil may be different from one separator to another one and consequently the mean residence time too, which will affect the efficiency of the process. Moreover, since the operating conditions are different (pressure, temperature, flow rate), the flow behavior may also be different. This will also influence the efficiency of the separation, the deposition of the sediments and the formation of the foam.

TRACER EXPERIMENTS AND IN-TERPRETATION OF THE RTD

Description of the Tracer Experiments

The nuclear technology allows carrying out these evaluations without stopping the process and the production of crude oil. Gamma radiotracer can be injected at the inlet of the studied process and continuously followed from the outside of pipes and separators with several detectors. The radiotracer used in the study was ^{113m}In, probably in the form of chloride complexes of the form ^{113m}In $(Cl)x^{3-}x$. It is obtained using a radionuclide generator of the kind $^{113}Sn/^{113m}In$. ^{113m}In has a half-life of 90 min, which is long enough to carry out the measurement that lasts less than 10 min. The maximum activity used in these experiments was 8 mCi ($\approx$0.3 MBq). The tracer injection (10 mL) was done against pressure using a nitrogen carboy as a pumping element. Under the present experimental conditions, the ^{113m}In in the eluent 0.05 N HCl acts as a marker of water in the crude oil, in emulsion form. This fact makes that the ^{113m}In in water follows the behavior of the movement, as physical marker, of the crude oil itself. The formation of small drops of ^{113m}In emulsion in the water is favored with the fall of pressure [3]. The detection system (FORCE Technology) was composed of several hermetic detectors (IC-2N2-WT-50m) connected with long polyurethane shielded wires to a compact processing data unit (IC-GDP) at which was coupled a computer RS-232 communication port. The hermetical detectors are composed of 2″ × 2″ NaI(TI) scintillation crystals. A typical radiotracer experiment is shown in Fig. 2.

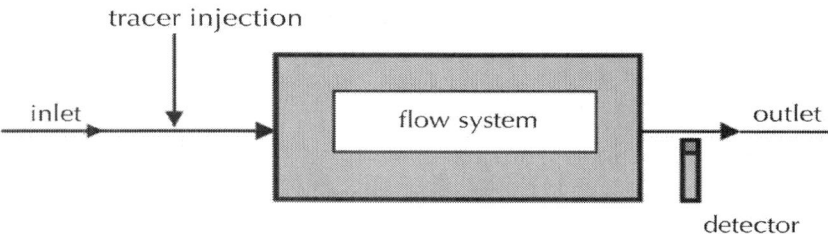

Figure 2: Typical radiotracer experiment in a simple flow system.

The radiotracer emitting γ radiation has been injected as a pulse function at the inlet of the studied system with a pressured injection system. The scintillation probe, placed at the outlet, detects the gamma radiation as a time function. Since the counting rate is directly proportional to the concentration of the radiotracer in the flowing stream where the probe is placed, the outlet curve gives the residence time distribution function (RTD). However, the technical difficulty to have a perfectly collimated beam causes errors on the RTD determination [10]. Tracer experiments have been conducted one by one in the three separators of the battery. The signals have been continuously recorded at the entrance of the separator, the gas outlet and the crude oil outlet. With the limitations mentioned above on the lack of proper detection collimation, the radioactive background has been properly removed from the original data, and then the curves have been normalized. Filtering of the curves for the determination of RTD using radiotracers is still subject of unlimited and contradictory scientific talk. We preferred to avoid this treatment which may result in a loss of information [5].

Residence Time Distribution Background

The residence time distribution (RTD) is a chemical engineering concept introduced by Danckverts [1]. It has been described in a multitude of scientific papers [2] and applied for various industrial processes. Supposing that the injection of tracer is made instantaneously at t = 0 (Dirac impulse), if C(t) is the curve representing the concentration of tracer at the outlet of the reactor, the fraction of molecule that remains in the studied system can be estimated at any value of time:

$$E(t) = \frac{C(t)}{\int_0^\infty C(t)\, dt}$$

(1)

(for a constant flow rate Q and uniform speed of fluid in the inlet and outlet sections)

E(t) is, by definition (see Eq. (1)), the impulse response (or RTD) of the system. It is easily obtained by dividing the concentration C(t) by the total surface under the curve. In the case of a non-instantaneous injection, whose function is in the form of x(t), the curve obtained on outlet is equal to the product of the convolution of x(t) by the impulse response E(t):

$$y(t) = \int_0^t E(u)x(t - u)\, du$$

(2)

This concept allows obtaining data information about hydrodynamic characteristics and troubleshooting [7]. The first moment of the experimental RTD gives the mean residence time of the material inside the studied process. The second moment leads to estimate of the dispersion, which is representative of the flow behavior. The accuracy of these parameters has been described in detail by Stegowski [9].

The precision on the third moment is rarely enough to allow proper interpretation. Better understanding of complex processes can be obtained by simulation of the experimental RTD using compartment models [7]. Depending on the complexity of the system fuzzy logic methods allows proposing multiple parameters model with a good confidence [6].

INTERPRETATION OF THE RESULTS

Crude Oil Separation Process

The mass balance in the process can be estimated by using process simulation software package (PRO II). However, the ratio of crude oil/gas contained in the separator is unknown. The first interpretation consists of calculating the mean residence time of the crude oil using the RTD distribution curves and, knowing the flow rate it is possible to estimate the volume of crude oil V_{oil} inside each separator.

The inlet signal has been recorded for each experiment. Fig. 3 shows that the signal can be considered as a Dirac impulse. The outlet signal gives a direct access to the residence time distribution and the simulation can be done easily without convolution.

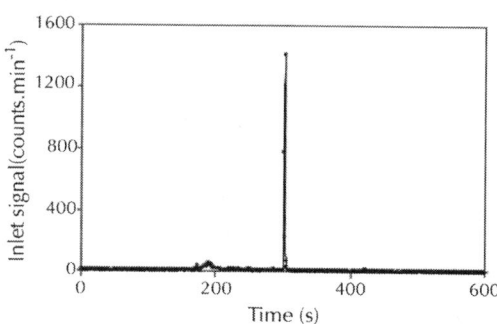

Figure 3: Typical raw inlet signal obtained during the measurements.

Although the predicted mass balance can be calculated by process software package (PRO II, for example), it is difficult to measure the internal volume occupied by the crude oil. The experimental RTD can already provide this key information by calculation of the first moment of the RTD. Table 2 lists the mean residence time (τ_{exp}) of the various separators and the estimated volume of crude (V_{oil}) inside each separator. This volume is calculated from Eq. (3) in which Q_{oil} is the flow rate of crude oil estimated by process software package PRO II:

$$V_{oil} = \tau_{exp} Q_{oil}$$

(1)

The mean residence times in the first and in the third separator have the same order of magnitude whereas it is nearly three time lower in the second one.

Table 2: Mean residence times of crude oil and volume of crude oil in the separators obtained by experiments and simulations

	Pressure (Pa)	Experi-ment, τ_{exp} (s)	Simula-tion, τ_{simul} (s)	V_{oil} (m³)	V_{oil}/V_{total} (%)
Separator No. 1	8.27×10^6	117	117	3	6.6
Separator No. 2	3.45×10^6	45	45	1.8	4
Separator No. 3	4.14×10^5	126	123	4.6	10.2

The percentages of the volume occupied by the crude oil in the total volume (V_{total} = 45.3 m³) of the separator (V_{oil}/V_{total}) vary from 4 to 10%. The possible explanation is not straightforward even if the operating pressure and the level of foam in the separator play certainly important roles. This result may confirm the possible deposition of sediment, which is more important in the two first separators as it has been estimated by preliminary gamma scanning experiments. Because of the sediment bed, the crude oil that is flowing through an evacuation at the bottom of the separator (see point E in Fig. 4) occupied a smaller volume.

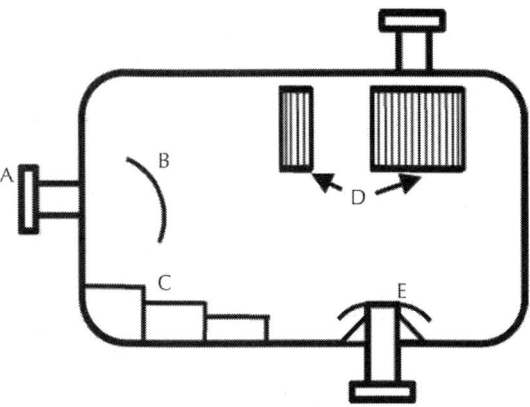

Figure 4: Scheme of the crude oil/gas separator. (A) Crude oil inlet; (B) deflector; (C) dissipation table; (D) filter; (E) crude oil outlet.

Modeling the flow behavior in the separators leads to more detailed information about the process. Since the three separators are similar, the proposed model should be the same for each one. Only the parameters are supposed to change with the operating conditions. Fig. 4 shows the scheme of the separator. At the inlet, the deflector will induce an important mixing due to the impact of the turbulent inlet flow. The dissipation table breaks the turbulence. Fig. 5, Fig. 6 and Fig. 7 show the good agreement obtained between experimental and simulated RTD using the chosen model described in Fig. 8. The model assumes that the flow is composed of a relatively well-mixed part due to the deflector located at the entrance and represented by a few perfect mixers in series. The second part of the model is a plug flow part due to the high viscosity of the crude oil. The precision of the parameters can be estimated to 10% due to the noise of the curve. Nevertheless, the multiple agreements for the three different experiments obtained with the same model are reasonable. When changing the value of one of the different parameters by more than 10%, the simulated curves strongly deviate from the experimental one. Fig. 9 shows the evolution of the number of cells in series with the operating pressure. Fig. 10 shows the percentage of the plug flow versus the operating pressure. The total volume of the crude oil inside the separator is linked to the parameters of the model according to Eq. (4):

$$V_{\text{oil}} = V_{\text{pmc}} + V_{\text{pf}} = Q_{\text{oil}}(\tau_{\text{pmc}} + \tau_{\text{pf}})$$

(4)

where V_{pmc} and V_{pf} are, respectively, the volume of the well-mixed part and the volume of the plug flow part. τ_{pmc} and τ_{pf} are the mean residence time, respectively, in the perfect mixing cells in series part and in the plug flow part.

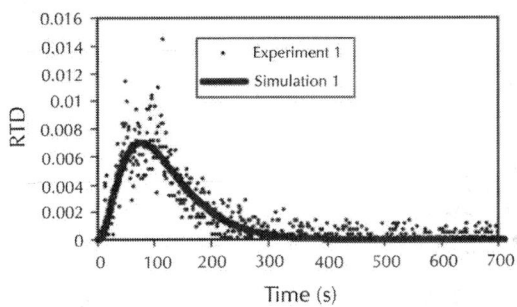

Figure 5: Comparison between experimental and simulated RTD for the first separator (pressure = 8.3×10^6 Pa).

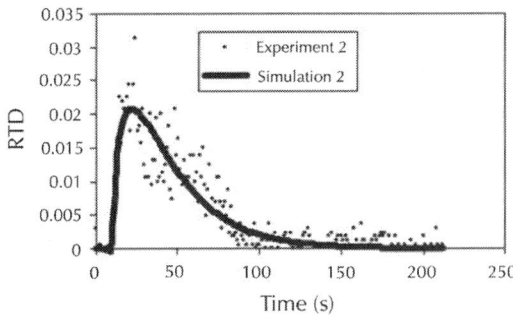

Figure 6: Comparison between experimental and simulated RTD for the second separator (pressure = 3.45×10^7 Pa).

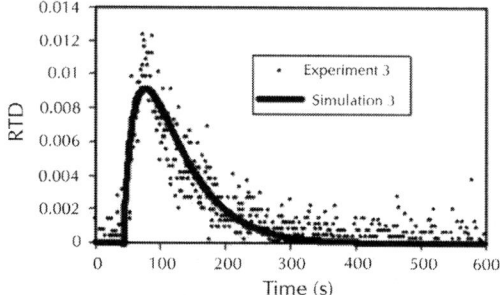

Figure 7: Comparison between experimental and simulated RTD for the third separator (pressure = 4.14×10^5 Pa).

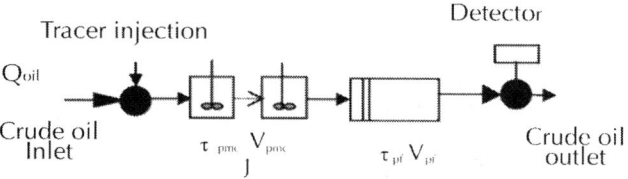

Figure 8: Schematic representation of the chosen model and the associated parameters.

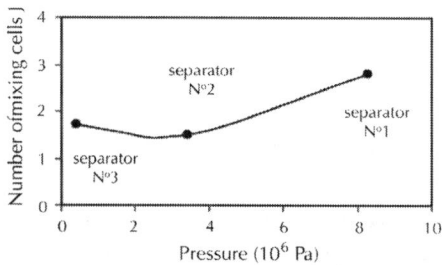

Figure 9: Number of mixing cells in series vs. operating pressures (i.e. for the three separators).

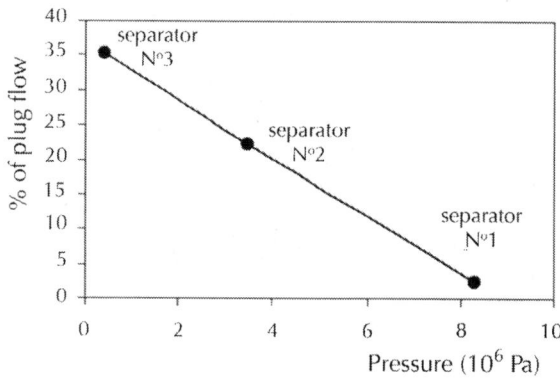

Figure 10: Percentage of plug flow fraction vs. operating pressure for the three different separators.

The plug flow part strongly decreases when the pressure decreases. This is due to the strong impact on the deflector that leads to a strong dispersion of the flow inside the entire separator. When the pressure decreases, the plug flow part increase up to 35% of the total volume. The number of mixing cells in the mixing part is slightly affected by the pressure under 4×10^6 Pa. In this pressure range (separators 2 and 3), the mixing induced by the deflector is very important since the number of mixing cells is close to 1 (around 1.5–1.7). For higher pressures, the part of the plug flow is very small and the mixing induced by the deflector affects the entire volume of the crude oil with a dispersion equivalent to 2.3 perfect mixing cells in series.

Gas Separation Process

As it has been indicated in the introduction, the tolerance for the concentration of oil in gas is zero. In order to control this aspect of the separation, the tracer concentration has been measured at the gas outlet of the three separators. Fig. 11, Fig. 12 and Fig. 13 show, respectively, the response of the detector located at the gas outlets of the three separators. For the two separators working at the higher pressure, the detectors have been collimated. The obtained signals (Fig. 11 and Fig. 12) intensity corresponding to the residual background and after the I injection. No evidence of traces of Indium has been detected. This proves only that there is no crucial malfunction in the gas separation process since the intensity (counting statistics) in the outlet curves is too low to conclude strongly on quantity of liquid carry-over with the gas and to assume that the tolerance is respected. A low concentration of oil will lead to a signal lower than the noise of the curve. For the last separator operating at a lower pressure, the detector has been used without collimator to increase the volume seen by the detector and consequently the intensity of the signal. A small increase of the signal has been detected just after the injection as shown in Fig. 13. The count rate (cpm) is significant enough but it is not evidence whereas the detector reacts to the activity inside the separator or to a low concentration of oil in the gas outlet. The proposed method is sufficient enough to detect a medium efficiency of the gas separation but not to control the strict tolerance requests by economic constraints of the oil production. Nevertheless, the obtained results may be considered as background information to go deeper in this research field. It should be pointed out that other possibilities have been studied and eliminated. One of them is to install a cyclone at the gas outlet and to control the cumulative accumulation of crude oil. However, this cyclone will induces a high pressure drop in the system, which is not acceptable for the exploitants. The solution may be found by using a tracer labeled with a lower-energy radionuclide and a higher activity, well-collimated detectors with sufficient shielding and a random inlet signal interpreted through the cross-correlation method. This application is not straightforward since even if this method has been used successfully for different industrial applications [8], the sensitivity requested in the presented example is very high. Recent published work [4] shows that the online measurements of liquid carry-over from scrubbers or separator using radioactive tracers requires suitable tracers which should be as representative as possible of the studied liquid phase. The authors showed that it is necessary to dissolve and inject the tracer with the same liquid, taken from the process. Nevertheless that the uncertainties in these measurements remains still high up to 14%, the tracer method seems to has the potential to become powerful tool for liquid carry-over measurements.

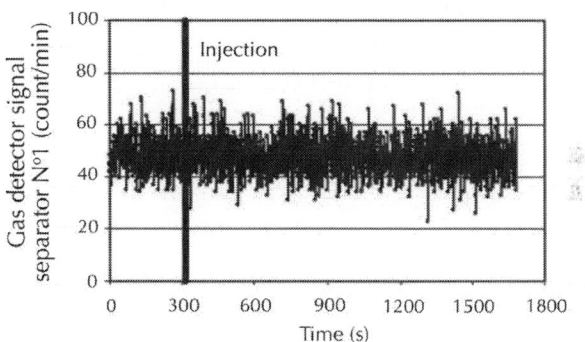

Figure 11: Experimental detector response at the gas outlet of the first separator (pressure = 8.37×10^6 Pa).

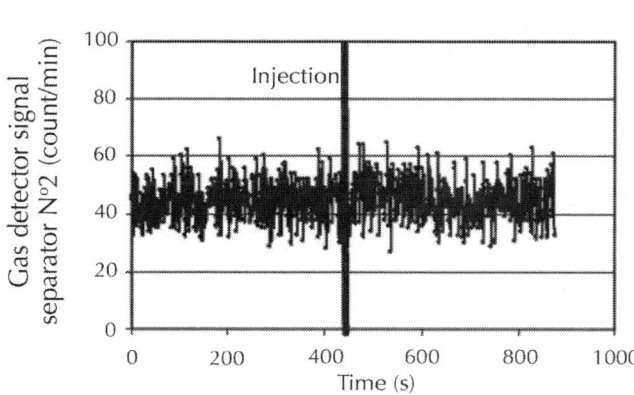

Figure 12: Experimental detector response at the gas outlet of the second separator (pressure = 3.45×10^6 Pa).

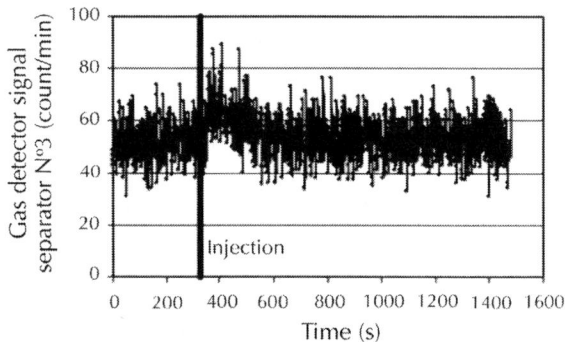

Figure 13: Experimental detector response at the gas outlet of the third separator (pressure = 4.14×10^5 Pa).

CONCLUSIONS

A radiotracer technique has been used to study the flow behavior in a battery of industrial crude oil/gas separators in oil industry. The battery is composed of three separators operating at different pressures. The real volume occupied by the crude oil has been determined by simple estimation of the first moments. These volumes varied from 4 to 10% of the geometrical ones. This difference confirms partially the deposition of sediments in the first separators of the battery as suspected by the operators. The obtained RTD curves have been simulated by a model composed of a few mixing cells in series representing the effect of the deflector located at the entrance and a plug flow part due to the high viscosity of the crude. The number of tanks in series remains constant around 1.6 for pressure lower than 4×10^6 Pa. The plug flow percentage is strongly depending on the pressure. It starts from 35% for low pressures to nearly 0 for high pressures. For high pressures, the turbulence induced by the deflector affects the entire volume of the crude oil with dispersion equivalent to 2.8 mixing cells in series. The analyses of gas separation shows that the impulse radiotracer method allows to detect non-negligible malfunction but it is not sensitive enough to control the zero tolerance of concentration of oil. The next step of this study is to compare tracer experiments presented in this paper with local neutron scanning of the separator. This method will provide a local estimation of the different level of sediments, crude oil and foam. We expect to compare the volume of crude oil obtained by the two methods and to compare the level of sediments with the proposed model.

ACKNOWLEDGMENTS

The International Atomic Energy Agency, projects ARCAL LXI, RLA/8/0281801, whose help is greatly acknowledged, supported this work. We would like to thank Mr. Thereska (IAEA) and Mr. Jin (IAEA) for their assistances during this project. We are very grateful to Mr. Mario Cano for his help during the experiments.

REFERENCES

1.   P.V. Danckverts, Continuous flow systems: distribution of residence times, Chem. Eng. Sci. 2 (1) (1953) 1–13.

2.   M.P. Dudukovic, Tracer methods in chemical reactors. Techniques and applications, in: Hugo I., De Lasa (Eds.), Chemical Reactor Design and Technology. NATO ASI Series E, Applied Sciences, 110, 1986.

3.   A. Finborud, M. Faucher, E. Sellman, Proceedings of the SPE Annual Technical Conference and Exhibition, Houston, TX, USA, October, 1999.

4.   A. Haugan, S. Hassfjell, A. Finborud, Online measurements of liquid carry-over from scrubbers using radioactive tracers, in: Proceedings of the Tracers and Tracing Methods, TRACER 3, Ciechocienek, 22–24 June, Institute of Nuclear Chemistry and Technology, Warszawa, 2004, pp. 293–298.

5.   J.-P. Leclerc, S. Claudel, H.G. Lintz, O. Potier, B. Antoine, Theoretical interpretation of residence time distribution

measurements in industrials processes: oil and gas science and technology, Rev. Inst. Franc͵ais du Petrole 55 (2) (2000) 1–12. ´

6.   S. Claudel, C. Fonteix, J.P. Leclerc, H.G. Lintz, Application of the possibility theory to the compartment modelling of flow pattern in complex processes, Chem. Eng. Sci. 58 (2003) 4005–4016.

7.   O. Levenspiel, Chemical Reaction Engineering, 3rd ed., Wiley, New York, 1999.

8.   L. Petryka, L. Furman, K. Przewlocki, Z. Stegowski, Radioisotope investigation of copper ore dressing processes, Nucl. Geophys. 7 (2) (1993) 313.

9.   Z. Stegowski, Accuracy of residence time distribution function parameters, Nucl. Geophys. 7 (2) (1993) 335.

10.   J. Thyn, R. Zitny, J. Kluson, T. Cechak, Analysis and Diagnostics of Industrial Processes by Radiotracers and Radioisotope Sealed Sources, Vydavatelstvi CVUT, Praha, 2000.

Chapter 3

Experimental Study of Particle Flow in a Gas–Solid Separator with Baffles Using PDPA

L. Du, J. Zh. Yao, and W.G. Lin

Multi-phase Reaction Laboratory, Institute of Process Engineering, Chinese Academy of Sciences, Beijing 100080, PR China

ABSTRACT

Particle flow behavior in a gas–solid separator with a guide baffle and separation baffles developed by the Institute of Process Engineering (IPE) for cocurrent down-flow reactors has been studied with a

Phase Doppler Particle Analyzer (PDPA), which allows to measure the particle size and velocity simultaneously. Profiles of particle velocities, particle size and particle number density at different heights in the separator were obtained. Analysis of the velocity profile indicates that the particle inertia has a significant effect on the behavior of particles in different size groups: the trajectories of the smaller particles are more influenced by the gas stream in relation to the larger, heavier, particles. Experimental results reveal that the guide baffle together with the separation baffles plays a role of guidance as well as particle concentrators. Particles entrained by the gas flowing upward along the top side of the separation baffle may mainly come from the regions far from the guide baffle surface. Small particles are gradually dispersed toward the periphery of the two-phase flow field after being ejected from the rectangular nozzle, giving rise to a bimodal behavior of particle size distributions in the region far from the guide baffle.

INTRODUCTION

In the last decade, more and more attention have been paid to co-current down-flow circulating fluidized bed reactors, so called 'downer' reactors. These reactors found a wide application, including fluidized catalytic cracking processes [1], [2] and [3], ultra-pyrolysis of cellulose biomass [4], gasification and liquefaction of coal [5] and [6]. The downer reactors offer significant advantages over the risers and conventional fluidized bed reactors for fast reactions due to its much more uniform gas–solid flow pattern, shorter contact time, low axial dispersion, higher solids/gas loading ratio, and ability to operate at a higher temperature [7]. For fast reactions with only a fraction of a second of gas–solid reaction/residence time, the quick separation of gas and solids at the outlet of the downer is, therefore, essential to prevent the overreaction occurrence and to ensure good productivity. A cyclone, with the residence time of the order of 1–2 s, may not be feasible for a downer reactor. Non-traditional gas–solids separators, with low residence time and high separation efficiency, are required. A short residence time gas–solids separation device has been patented by Stone & Webster Engineering Corporation [8]. It is a U-turn inertial separator. Separation is effected by projecting solids by centrifugal force against a bed of solids as the gas phase makes a 180° directional change and solids only 90° change relative to the incoming stream. By formation of the one-quarter-curve bed of solids, erosion of the wall opposite the inlet of the separator is eliminated and a U-shaped 180° flow pattern of the gas stream is aided to establish. Although a very short residence time were claimed for the separator, the separation efficiency was not high enough due to significant gas entrainment.

For the same purpose, another gas–solid separator for downer reactors has been developed by Tsinghua University in the 1980s [9]. It is a simple inertial separator in which gas and solids suspension first pass through a specially designed nozzle and then impinge on a curved guiding plate with a gradually increasing radius.

Recently, the Institute of Process Engineering, Chinese Academy of Sciences, has proposed a novel gas–solid separator, the Combined Baffles Quick-Separation Device (CBQSD), which has been applied to flash pyrolysis of coal in a downer. Previous experimental results have shown that the CBQSD has some advantages, such as high separation efficiency, low-pressure loss, especially its relatively compact and symmetric structure [10]. It has also been found that separation performances (separation efficiency, pressure drop, etc.) largely depend on geometrical parameters. The optimum design and scale-up of the CBQSD requires a thorough understanding of gas–solid flow patterns. To achieve this understanding, detailed experimental measurements of the gas–solid flow provide a basic tool.

Many techniques have been used for characterizing multiphase flow patterns, including X-ray densitometer [11], -ray computed tomography [12] and [13], electrical capacitance tomography [14] and [15], Laser Doppler Anemometry [16] and Particle Image Velocimetry (PIV) [17], etc. In the present paper, the solid-phase hydrodynamics in the CBQSD was studied by using a Phase Doppler

Particle Analyzer (PDPA). The PDPA is a non-intrusive, laser-based measurement technique based on measuring the Doppler signal scattered by the particle as it cross the sample volume. It allows the simultaneous measurement of size and velocity of spherical particles, droplets, or bubbles. Until now, this measuring technique has been successfully used for particle measurements in a wide variety of situations, such as swirling flow [18] and [19] and gas-particle flow in a circulating fluidized bed [20].

The objective of this study is to understand the separation mechanism of the newly developed separator CBQSD and to investigate the solid-phase hydrodynamics under a given operating and geometrical conditions with a two-component PDPA to obtain particle size, velocity and number density.

EXPERIMENTAL DETAILS

Experiments were carried out in a setup consisting of a riser, a downer and a gas–solid separator, which is schematically shown in Fig. 1. An air compressor supplied air to a two-dimensional separator (CBQSD) via a riser with 32 mm inner diameter and a downer with 40 mm inner diameter. The riser and the downer had the same height of 3 m. The airflow rate was controlled by a valve and a flow meter. Glass beads were loaded into the airflow by a feeding system consisting of a valve and a particle container. After gas–solid separation, the separated particles entered into a solids tank below the separator, while the unseparated particles were entrained out of the separator by the gas flow and were collected by bag-type collectors.

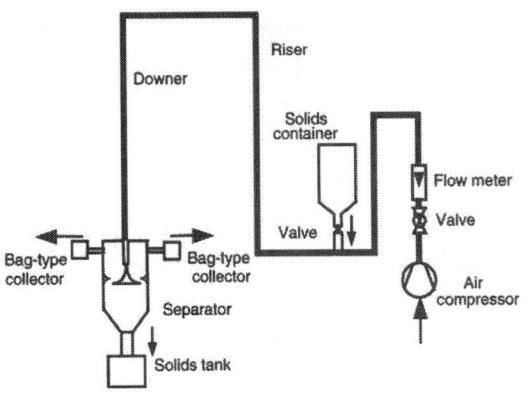

Figure 1: Schematic of the experimental apparatus.

The geometrical structure of the two dimensional separator is shown in Fig. 2a. The separator consists of three internal components: a rectangular nozzle (RN) which introduces the gas–solid flow into the separator, a guide baffle (GB) which is composed of two circular arcs in bilateral symmetry, and a separation baffle (SB) which is also composed of two circular arcs, SB-top and SB-bottom, but in up-down symmetry. Compared with the downer tube, the rectangular nozzle has a smaller cross-section. The separator is made of Plexiglas, which enables PDPA measurements possible.

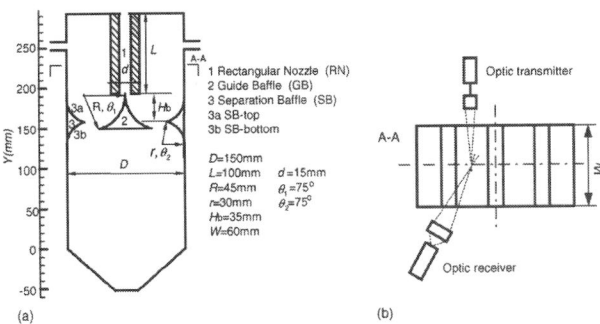

Figure 2: Geometrical configuration of the separator. (a) Frontal bi-dimensional scheme of the separator geometry and (b) scheme of the PDPA measuring system.

The particle size, velocity and number density in the separator were measured by an Aerometrics two-component PDPA. The principle of PDPA was described by Bachalo and Houser [21]. The arrangement of the measuring system is shown in Fig. 2b.

In the measuring system, particle size is determined by using green beams with a wavelength of 0.5145 µm, which also give the first component of particle velocity, i.e., the vertical velocity u, while the 0.488 µm blue beams provide the second orthogonal component of particle velocity, i.e., the transversal velocity v. The selected optical components are a transmitter lens, a receiver lens with f = 500 mm and receiver back lens with f = 238.6 mm. The combination gives an effective detecting range of 0.5–142 µm of the particle size. For glass beads the PDPA was configured in the 22° off-axis forward scatter for this study.

Experimental conditions are listed in Table 1 and the particle-size distribution measured at one point of the nozzle outlet by a PDPA is also given in Fig. 3.

Table 1: Flow conditions

Air flow	
Mean gas velocity (at the nozzle outlet) (m/s)	3.7
Reynolds number (obtained with nozzle D_h = 24 mm)	6000
Particles	
Particle arithmetic mean diameter (µm)	67
Particle material density (kg/m³)	2500
Particle mass flow rate (g/s)	5.33
Particle loading	1.31

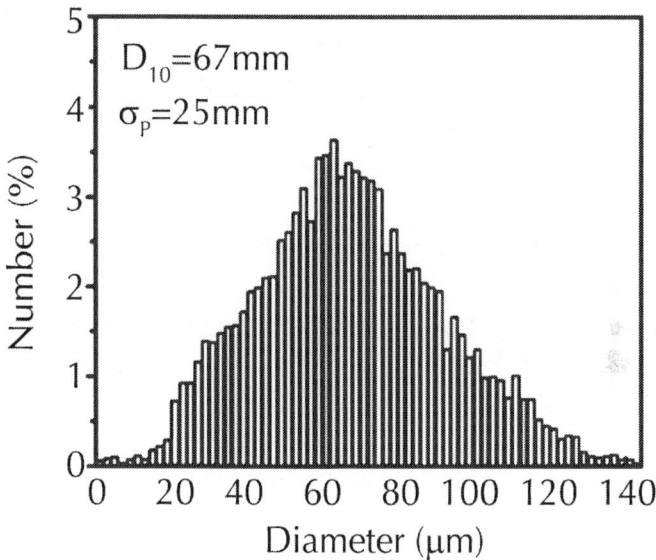

Figure 3: Particle size distribution.

Particle behavior near the guide baffle and the separation baffles were of our major interest. Thus, measurements in this study were focused on those regions. Experiments were performed with transversal scans at nine different heights, Y = 190, 180, 175, 170, 165, 160, 155, 150, 140 mm (Fig. 2a), to obtain particle velocity, particle size and particle number density in the separator. At each height, the distance between measurement points was kept 1 or 2 mm. The measurements started at the point as close to the surface of the GB and SB as possible. When the particle number density was very low at the point far from the surface, the PDPA measurements were stopped. In order to achieve reasonably accurate local mean parameters (velocity, size, etc.) of the particle behaviors, twenty thousand particles were measured at each location.

EXPERIMENTAL RESULTS AND DISCUSSION

Particle Velocity

Particle Mean Velocity Vectors in the Separator

Fig. 4 plots particle mean velocity vectors that indicate the motion of particles in the separator. The two components of the particle mean velocity vectors, the vertical and horizontal, represent particle mean vertical and transversal velocity, respectively, averaged over all particle sizes. The positive directions of particle vertical and transversal velocity are defined to be vertically upward and horizontally right, respectively. In Fig. 4, it is observed that particles flow vertically downward from the inlet rectangular nozzle along to the guide baffle. At the end of this baffle, some particles move downward due to the inertia effect, being eventually settled down. Other particles flow upward along the SB-top being entrained out of the separator by the gas flow. Conclusions could be drawn that in the gas–solid separation process the guide baffle and the separation baffles act as the guides, which change and direct the motion of the particles in the separator.

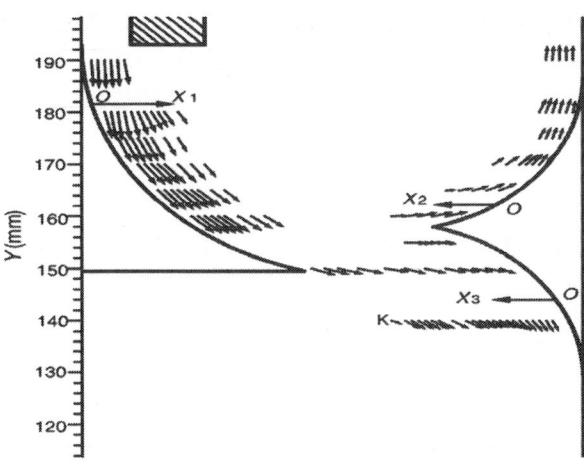

Figure 4: Particle mean velocity vectors in the separator.

For the convenience of further study, different horizontal co-ordinate OX_1, OX_2, OX_3 are adopted for the guide baffle, and top and bottom regions of the separation baffles, respectively, as shown in Fig. 4. The origins are set at the intersection of horizontal co-ordinate and baffle surfaces. So, the co-ordinate at one point near the guide baffle surface, X_1 of the OX_1-axis, for instance, represents the horizontal distance from the point to the guide baffle surface.

Particle Velocity near the Guide Baffle

Fig. 5 shows the profiles of particle mean vertical U_p (a) and transversal V_p (b) velocities averaged over all particle sizes along the guide baffle for heights from Y = 190 to 150 mm, respectively. At all heights the vertical velocity is high close to the guide baffle; whereas lower in the region far from the guide baffle. Also, noticeable is that the particles exhibit a decrease in vertical velocity along the guide baffle from Y = 190 to 150 mm but variations of the particle vertical velocity between two neighboring vertical locations are different: largest between Y = 160 and 150 mm, while smallest between Y = 190 and 180 mm. However, for the transversal velocity profiles (Fig. 5a) this is not the case at all. The profiles are quite different between each other. In the transversal direction, the particles are accelerated along the guide baffle. There is a lowest value of transversal velocity in the region close to the guide baffle at Y = 190 and 180 mm, quite different from the profiles at Y = 170 and 160 mm where a highest value is found in the region close to the guide baffle. The reason for this behavior is that after being ejected downward from the nozzle with almost zero transversal velocity, the particles which are located in the region closest to the guide baffle will first collide with the guide baffle and then rebound, attaining a high transversal velocity value. The particles located far from the guide baffle increase gradually their transversal velocity due to the drag force exerted by the gas flow. As a result, a lowest value of transversal velocity is obtained in the region close to the guide baffle at Y = 190 and 180 mm. At the end of the guide baffle, i.e., Y = 150 mm, an almost uniform distribution of transversal velocity is seen.

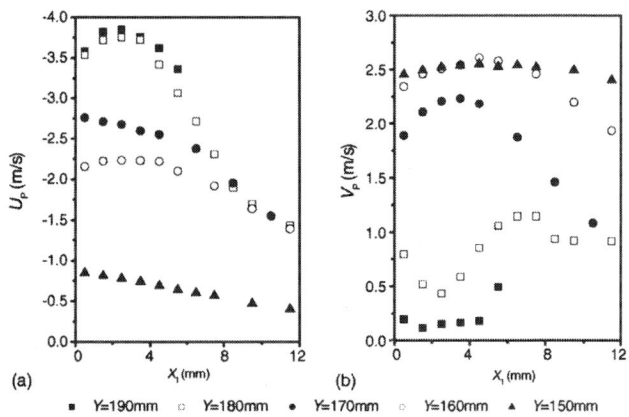

Figure 5: Profiles of particle mean vertical velocity (a) and mean transversal velocity (b) along the guide baffle from Y = 190 to 150 mm.

To investigate the behavior of different size particles, the data of particle size and velocity are reprocessed after the measurements. The particle size distribution in the range from 30 to 100 μm is resolved by seven classes of 10 μm width. In Fig. 6, the mean vertical velocity is shown for four particle size classes (35, 55, 75 and 95 µm) of 10 µm width at five heights near the guide baffle. It is seen that at Y = 190 mm which is very close to the outlet of the rectangular nozzle the smaller the particles are, the higher the vertical velocity is, probably caused by faster acceleration for the smaller particles due to their lower inertia in the 100 mm rectangular nozzle.

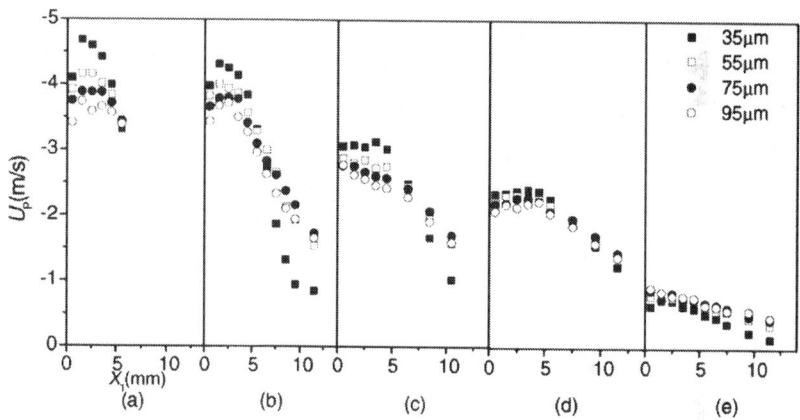

Figure 6: Distributions of mean vertical velocity of different size particles at (a) Y = 190 mm, (b) Y = 180 mm, (c) Y = 170 mm, (d) Y = 160 mm and (e) Y = 150 mm near the guide baffle.

From this location until Y = 160 mm, the smaller particles still have larger vertical velocity in the region close to the guide baffle. However, the differences of vertical velocity among different size particles are gradually decreasing. The small particles are decelerated so fast that when they reach the end of the guide baffle, i.e., Y = 150 mm, their vertical velocities are below large particles' vertical velocities, as shown in Fig. 6e. It is also observed that in the region far from the guide baffle at Y = 180 and 170 mm the smaller particles (35 µm) have lower velocity. A possible reason for above behavior of smaller particles is given as the following.

One possible explanation is due to the change in the gas velocity. Observing the geometry of the guide baffle it is reasonable to assume that the vertical component of the gas vector velocity will decrease with the co-ordinate Y diminution. Considering that the smaller particles have a lower inertia, it can be assumed that these particles will be more influenced by the gas flow, than the larger ones. This behavior will lead to a more notably mean diminution of the smaller particles vertical velocity in comparison with the larger particles.

Particle Velocity near the SB-Top and SB-Bottom

Fig. 7 shows the profiles of vertical mean velocity (a) and transversal mean velocity (b) averaged over all particle sizes along the SB-top from Y = 160 to 190 mm. It can be seen from Fig. 7 that when going upward along the SB-top surface, the particles reveal an increase in vertical velocity but a decrease in transversal velocity.

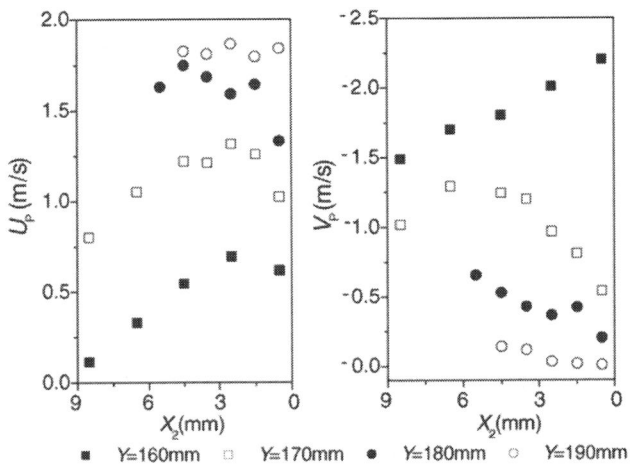

Figure 7: Profiles of particle mean vertical velocity (a) and mean transversal velocity (b) along the SB-top from Y = 160 to 190 mm.

Fig. 8 shows the distributions of mean vertical velocity at (a) Y = 160 mm, (b) Y = 170 mm, (c) Y = 180 mm, (d) Y = 190 mm near the SB-top for five particle size classes, namely, 15, 35, 55, 75, 95 μm. As shown inFig. 7a, all the particles are accelerated along the

SB-top. In Fig. 8, it is observed that at each height smaller particles generally exhibit higher vertical velocity. This behavior is probably caused by its lower inertia, being more accelerated due to the drag force effect of the gas flow. The profiles of mean vertical velocity of different size particles near the separation baffle-top together with those near the guide baffle shown in Fig. 6 manifest the significant effects of inertia on the behavior of different size particles: smaller particles more rapidly respond to the change of the gas flow.

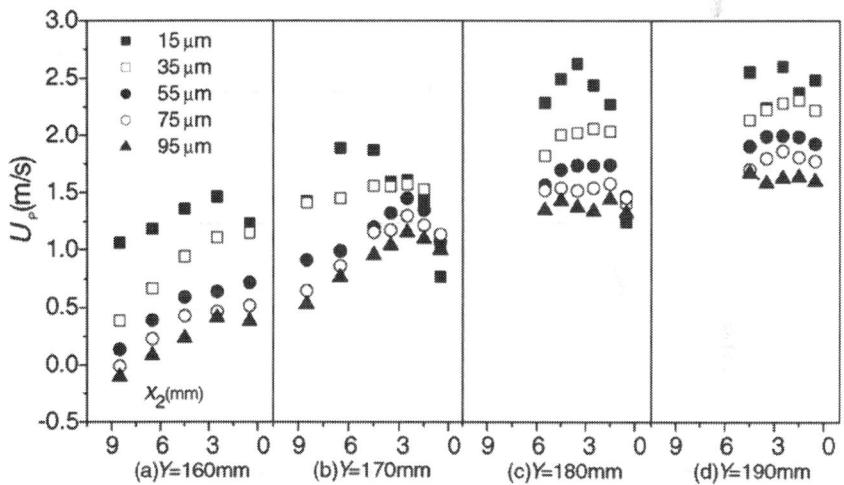

Figure 8: Distributions of mean vertical velocity of different size particles near the SB-top at (a) Y = 160 mm, (b) Y = 170 mm, (c) Y = 180 mm, (d) Y = 190 mm.

Fig. 9a illustrates the distribution of particle vertical velocity at Y = 155 mm near the SB-bottom for four size classes (35, 55, 75 and 95 μm). It should be noticed that small particles (i.e., 35, 55 μm) show a positive value of vertical velocity at Y = 155 mm near the SB-bottom, which means that the gas is flowing upward in this region, entraining small particles upward. However, larger particles show a negative vertical velocity atY = 155 mm near the SB-bottom, where the velocity of gas flow may be too low to entrain the larger particles.

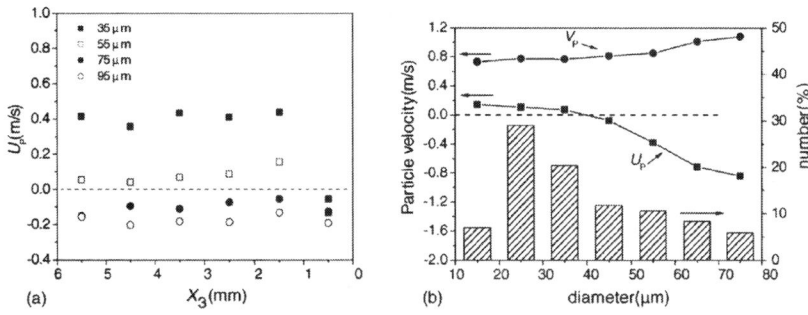

Figure 9: (a) Particle mean velocity distributions near the SB-bottom at Y = 155 mm for 35, 55, 75, 95 μm particles (b) Particle mean velocity together with particle size distributions at a point K (X_3 = 24.5 mm, Y = 140 mm) shown in Fig. 4.

The behavior of the gas–solid flow at a point K near the SB-bottom (X_3 = 24.5 mm, Y = 140 mm) shown inFig. 4 is shown in Fig. 9b. In this figure, is shown the particle mean velocity for different size classes from 10 to 80 μm associated with the particle size distribution. It is observed that at the point K the concentration of smaller particles is greater and larger particles have higher vertical and transversal (in module) velocities. The very low number density of the smaller particles with the mean diameter equal to 15 μm is in relation to the location of analyzed point K. If the analyzed point K is located much farther away from the surface of SB-bottom, the number density of 15 μm particles may be much higher. The smaller particles (i.e., 15, 25, 35 μm) exhibit a positive value of mean vertical velocity, which suggests that there can exist a gas recirculation in the region far from the SB-bottom.

Particle Size

Fig. 10 shows the distributions of the arithmetic mean diameter D_{10} near the guide baffle at heightsY = 190 mm (a), Y = 180 mm (b), Y = 170 mm (c), Y = 160 mm (d), Y = 150 mm (e), near the SB-top at the height Y = 160 mm (f), and near the SB-bottom at the height Y = 140 mm (g). It can be observed that the arithmetic mean diameter

D_{10} close to the guide baffle is generally larger than that far from the guide baffle at Y = 190 mm (a), Y = 180 mm (b), Y = 170 mm (c), Y = 160 mm (d) and Y = 150 mm (e), respectively. This implies that a larger number percent of small particles may be found in the region far from the guide baffle. Also, noticeable is that there is a significant increase in the arithmetic mean diameter near the guide baffle from Y = 160 mm (d) to Y = 150 mm (e), however, there is a significant decrease from Y = 160 mm (d) near the guide baffle to Y = 160 mm (f) near the SB-top. This suggests that when particles go along the guide baffle from Y = 160 mm to Y = 150 mm, some small particles are entrained by the gas flow to the SB-top. At the height Y = 140 mm (g), near the SB-bottom, the arithmetic mean diameter close to the SB-bottom is much larger than that far from the SB-bottom where small particles, accounting for larger fractions, may be found.

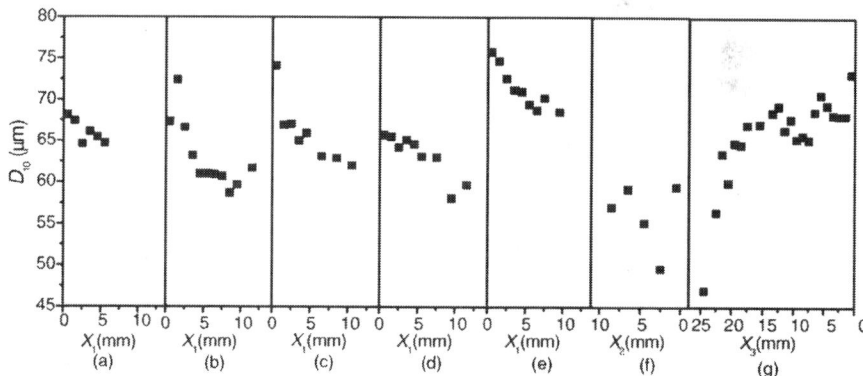

Figure 10: Distributions of the arithmetic mean diameter near the guide baffle at heights Y = 190 mm (a), Y = 180 mm (b), Y = 170 mm (c), Y = 160 mm (d), Y = 150 mm (e), near the SB-top at the height Y = 160 mm (f) and near the SB-bottom at the height Y = 140 mm (g).

In an attempt to further explain the foregoing results, the evolutions of particle size distributions are depicted in Fig. 11 at three transversal points (three columns) for each of five heights (five rows), namely, near the guide baffle: Y = 190 mm (the first row),

Y = 180 mm (the second row), Y = 160 mm (the third row), near the SB-top Y = 160 mm (the fourth row), near the SB-bottom Y = 140 mm (the fifth row). The first row shows that particles near the guide baffle at Y = 190 mm (which is very close to the outlet of the rectangular nozzle) are mono-modal distributions with peaks at about 70 μm. It should be noticed that at Y = 180 mm and Y = 160 mm near the guide baffle the particle size distributions vary gradually from a mono-modal distribution with a peak at 70 μm (left) to a bimodal one with two peaks approximately at 30 μm (left) and 70 μm (right), which means the fraction of small particles increases with the increase of the distance to the guide baffle. The variations of particle size distributions clearly indicate that small particles are gradually dispersed toward the periphery of the two-phase flow field after being ejected from the rectangular nozzle. At all three points for Y = 160 mm, near the SB-top (the fourth row), particles show bimodal distributions with two peaks approximately at 30 and 70 μm, probably relating to the bimodal behaviors mentioned above at locations far from the guide baffle for the heights Y = 160 and 180 mm. This means that some particles entrained by the gas flow at Y = 160 mm near the SB-top may come from the regions far from the guide baffle. For the height Y = 140 mm (the fifth row) near the SB-bottom particles show the same evolution behavior as at Y = 180 and 160 mm near the guide baffle: small particles are becoming more and more as X_3 increases. Especially at X_3 = 24.5 mm particles mainly consists of small ones which move upward as already indicated in Fig. 9b.

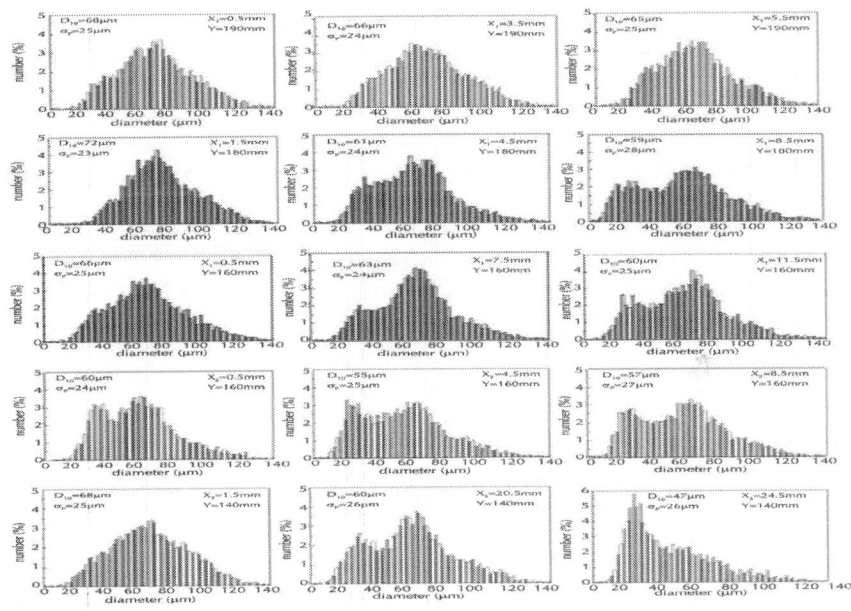

Figure 11: Evolutions of particle size distributions at three transversal positions (three columns) for each of five heights (five rows), namely, near the guide baffle: Y = 190 mm (the first row), Y = 180 mm (the second row), Y = 160 mm (the third row), near the SB-topY = 160 mm (the fourth row) and near the SB-bottom Y = 140 mm (the last row).

Particle Number Density

Fig. 12 illustrates the transversal distributions of particle number density at different heights in the separator CBQSD. It can be observed from Fig. 12a that near the guide baffle at Y = 190 mm particles are narrowly contained within the outlet of the rectangular nozzle. When particles reach the height ofY = 180 mm near the guide baffle, a wider number density distribution is found. Close to the guide baffle the particle concentration is high; however, the number density decays quickly as the distances to the guide baffle X_1 increases. So, are the distributions for heights at Y = 170, 160 and 150 mm near the guide baffle. The number density distributions near the guide baffle imply that after being ejected from the rectangular nozzle, although a few number of particles are dispersed into the regions far from the guide baffle, most of them are still concentrated close to the guide baffle due to inertia effects. It can also be seen from Fig. 12a that at Y = 140 mm near the SB-bottom with the distance to the SB-bottom X_3 increasing, the number density reduces sharply, which means that after particles leave the guide baffle at Y = 150 mm where particles are mainly concentrated close to the guide baffle, they gather very close to the SB-bottom again.

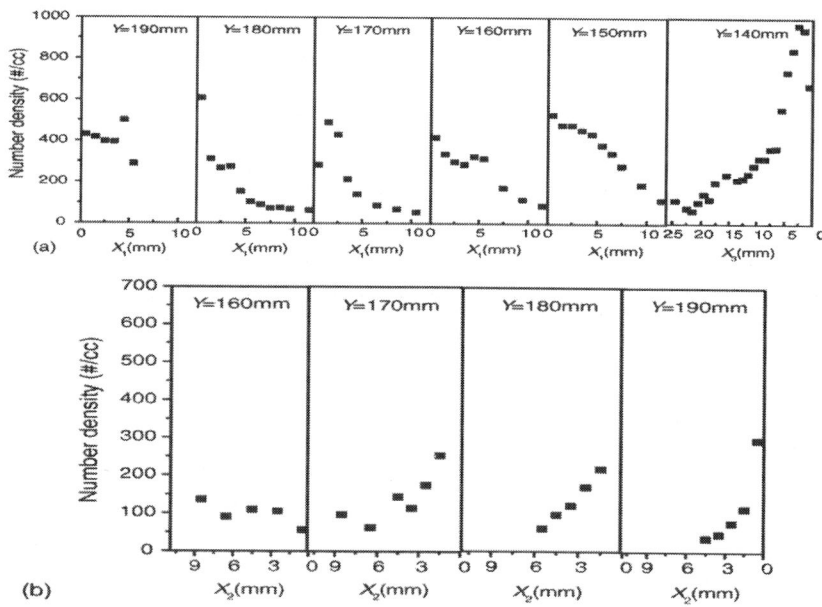

Figure 12: Particle number density distributions (a) near the guide baffle at Y = 190, 180, 170, 160, 150 mm and near the SB-bottom atY = 140 mm, (b) near the SB-top at Y = 160, 170, 180, 190 mm.

Fig. 12b shows the particle number density distributions near the SB-top at different heights. AtY = 160 mm, a slightly high particle number density is found far from the SB-top, while particle concentrations are relatively high close to the SB-top for each height of Y = 170, 180 and 190 mm. The reason is probably that particles, which move farther from the guide baffle at Y = 160 mm are more easily and largely entrained by the U-turn gas flow than those closer to the guide baffle. When going along the SB-top from Y = 170 to 190 mm, a large number of particles entrained by the upward gas flow concentrate again close to the SB-top due to the inertial effect.

From the above discussions, it can also be concluded that besides guidance, the guide baffle and the baffles, including the SB-top and the SB-bottom also play a role of particle concentrator, which make the particles gather close to their surfaces and are favorable to the gas–solid separation process.

CONCLUSIONS

The particle flow behavior in a gas–solid separator with a guide baffle and separation baffles developed by the Institute of Process Engineering (IPE) for co-current down-flow reactors have been studied with a Phase Doppler Particle Analyzer. Profiles of particle velocities, particle size and particle number density at different heights in the separator were obtained and analyzed.

When the mean vertical and transversal velocity profiles are concerned, the results obtained show that along the guide baffle the particles are decelerated in vertical mean velocity but accelerated in mean transversal velocity; on the other hand, along the SB-top the particles are accelerated in vertical velocity but decelerated in transversal velocity. The inertia of particles has a significant effect on the behavior of particles in different size groups, which means that smaller particles are more influenced by the drag force of the gas stream.

Results reveal that the guide baffle together with the separation baffles play a role of guidance as well as particle concentrators, which are very important to the gas–solid separation process. Particles entrained by the gas flow upward along the SB-top may mainly come from the regions far from the guide baffle. Small particles are gradually dispersed toward the periphery of the two-phase flow field after being ejected from the rectangular nozzle, giving rise to a bimodal behavior of particle size distributions in the region far from the guide baffle.

In the present paper, PDPA measurements were performed under a given operating and geometrical conditions. The influence of the baffles' geometrical form, the gas mass flow rate and turbulence on the particle hydrodynamics was not taken into consideration. In order to fully understand the separation mechanism, further study of the influence of these factors on the particle hydrodynamics using PDPA need to be carried out.

ACKNOWLEDGMENTS

The authors acknowledge with gratitude the financial support of the National Program of High Technology Research and Development (No. 2003AA514023 and No. 2001AA529010) and the National Natural Science Foundation of China (No. 20221603 and No. 90210034).

REFERENCES

1.   J.A. Talman, L. Reh, An experimental study of fluid catalytic cracking in a downer reactor, Chem. Eng. J. 84 (3) (2001) 517.

2.   A.G. Maadhah, M. Abul-Hamayel, A.M. Aitani, T. Ino, T. Okuhara, Down-flowing FCC reactor, Oil Gas J. 98 (33) (2000) 66.

3.   J.A. Talman, R. Geier, L. Reh, Development of a downer reactor for fluid catalytic cracking, Chem. Eng. Sci. 54 (13–14) (1999) 2123.

4.   B.A. Freel, R.G. Graham, M.A. Bergougnou, R.P. Overend, L.K. Mok, Kinetics of the fast pyrolysis (ultrapyrolysis) of cellulose in a fast fluidized bed reactor, AIChE Symp. Ser. 83 (1987) 105.

5.   Y.J. Kim, S.H. Lee, S.D. Kim, Coal gasification characteristics

in a downer reactor, Fuel 80 (13) (1915).

6. C.D. Oberg, A.Y. Falk, Coal liquefaction by flash hydropyrolysis, Coal Process Technol. 6 (1980) 159.

7. J.X. Zhu, Z.Q. Yu, Y. Jin, J.R. Grace, A. Issangya, Cocurrent down- flow circulating fluidized bed (downer) reactors-a state of the art review, Can. J. Chem. Eng. 73 (1995) 662.

8. R. J. Gartside, H. N. Woebcke, Low residence time gas–solid separation device and system, USA Patent: 4556541, 1985-12-03.

9. C.M. Qi, Z.Q. Yu, Y. Jin, X.L. Cui, X.X. Zhong, A novel inertial separator for gas–solid suspension in concurrent downflow, Pet. Refining 12 (1989) 51 (in Chinese).

10. S.G. Li, W.G. Lin, J. Zh. Yao, Novel separator for gas–solids cocurrent down-flow reactors, Chin. J. Process Eng. 2 (1) (2002) 12.

11. A. Miller, D. Gidaspow, Dense, vertical gas–solid flow in a pipe, AIChE J. 38 (11) (1992) 1801.

12. S. Roy, J.W. Chen, S.B. Kumar, M.H. Al-Dahhan, M.P. Dudukovic, Tomographic and particle tracking studies in a liquid–solid riser, Ind. Eng. Chem. Res. 3611 (1997) 4666.

13. T. Schiewe, K.E. Wirtha, O. Molerusa, K. Tuzlaa, A.K. Sharmab, J.C. Chenb, Measurements of solid concentration in a downward vertical gas–solid flow, AIChE J. 45 (5) (1999) 949.

14. A.J. Jaworski, T. Dyakowski, Application of electrical capacitance tomography for measurement of gas–solids flow characteristics in a pneumatic conveying system, Meas. Sci. Technol. 12 (8) (2001) 1109.

15. Zh.Y. Huang, B.L. Wang, H.Q. Li, Dynamic voidage measurements in a gas soild fluidized bed by electrical capacitance tomography, Chem. Eng. Commun. 190 (10) (2003) 1395.

16. Y.F. Zhang, A. Hamid, Dilute fluidized cracking catalyst particlesgas flow behavior in the riser of a circulating fluidized bed, Powder Technol. 84 (3) (1995) 221.

17. K. Miyazaki, G. Chen, F. Yamamoto, J. Ohta, Y. Murai, K. Horii, PIV measurement of particle motion in spiral gas–solid two-phase flow, Exp. Thermal Fluid Sci. 19 (4) (1999) 194.

18. M. Sommerfeld, H.H. Qiu, Detailed measurements in a swirling particulate two-phase flow by a phase-Doppler anemometer, Int. J. Heat Fluid Flow 12 (1) (1991) 20.

19. A. Brena de la Rosa, S.V. Sankar, G. Wang, W.D. Bachalo, Particle diagnostics and turbulence measurements in a confined isothermal liquid spray, J. Eng. Gas Turbines Power, Trans. ASME 115 (3) (1993) 499.

20. T. Van den Moortel, E. Azario, R. Santini, L. Tadrist, Experimental analysis of the gas-particle flow in a circulating fluidized bed using a phase Doppler particle analyzer, Chem. Eng. Sci. 53 (10) (1998) 1883.

21. W.D. Bachalo, M.J. Houser, Phase/Doppler spray analyzer for simultaneous measurements of drop size and velocity distributions, Opt. Eng. 23 (5) (1984) 583.

Chapter 4

Experimental Model Validation for N-Propyl Propionate Synthesis in a Reactive Distillation Column Coupled with a Liquid–Liquid Phase Separator

T. Keller[a], J. Muendges[a], A. Jantharasuk[a, b], C.A. Gónzalez-Rugerioa, H. Moritz[a, c], P. Kreis[a, d], and A. Górak[a]

[a]TU Dortmund University, Department of Biochemical and Chemical Engineering, Laboratory of Fluid Separations, Emil-Figge-Strasse 70, D-44227 Dortmund, Germany

[b]Center of Excellence in Catalysis and Catalytic Engineering, Department of Chemical Engineering, Faculty of Engineering, Chulalongkorn University, Bangkok 10330, Thailand

cBYK-Chemie GmbH, Abelstrasse 45, D-46483 Wesel, Germany

dEvonik Degussa GmbH, Paul-Baumann-Straße 1, D-45772 Marl, Germany

ABSTRACT

In the synthesis of some organic esters, reactive distillation coupled with a liquid–liquid phase separator is often used to increase the product purity or to recover the reactants. In this article, we present a comprehensive experimental and theoretical study on the heterogeneously catalysed synthesis of *n*-propyl propionate by reactive distillation and a subsequent liquid–liquid phase separator. The experiments were performed in a pilot-scale reactive distillation column. Data-reconciliation tests proved that the experimental results obtained comprise a complete, reliable set of composition and temperature profiles along the pilot-scale reactive distillation column and can be used for further model validation. A nonequilibrium-stage model was applied to predict the experimental results. Simulation studies demonstrated that the composition and temperature profiles in the rectifying section of the column were highly sensitive to the composition of the reflux stream entering the column. Deviations between the experimental and predicted composition profiles in the rectifying section were identified. An explanation for the deviations is given in this article.

INTRODUCTION

Reactive distillation (RD), a process in which chemical reaction and thermodynamic separation are integrated in a single apparatus, represents one of the best-known examples of process intensification (Harmsen, 2007 and Schoenmakers and Beßling, 2003). It allows for higher reactant conversion, product selectivity and energy savings with favourable investment and operating costs. Despite these advantages, RD has several constraints, such as complex design, difficult scale-up and advanced process control (Sundmacher and Kienle, 2003). Nevertheless, the concept of RD has been industrially applied for certain types of reactions, the most important being esterifications (Agreda et al., 1990), transesterifications (Steinigeweg and Gmehling, 2004) and etherifications (Sundmacher and Hoffmann, 1996), in which the maximum reactant conversion is limited by chemical equilibrium.

A reactive distillation column used for esterification reactions can be coupled with a liquid–liquid phase separator at the top of the column. There are two different reasons to use a liquid–liquid phase separator; these in turn depend upon whether the ester is a low- or high-boiling component in the reaction system. In the first case, the separator can ensure higher product purities by separating the organic ester from the aqueous phase. For example, Kloeker et al. (2003) reported the successful use of a liquid–liquid phase separator in the synthesis of ethyl acetate to further purify the distillate stream. The same reaction system was studied by Lai et al. (2008). They investigated different start-up procedures experimentally in the production of high purity ethyl acetate and demonstrated that initial holdup compositions in the column and in the liquid–liquid separator play an important role for an efficient start-up. Similar results were reported byForner et al. (2008) in the synthesis of n-propyl acetate. In the second case, the ester is a high-boiling component and leaves the RD column with the bottom stream. Here, a liquid–liquid phase split of the distillate stream enables the discharge of the aqueous phase and recycling of the organic phase, the latter mainly consisting of unconverted reactants, back to the RD column. In this context, Schmitt et al. (2004) presented a comprehensive study on the synthesis of n-hexyl acetate by reactive distillation coupled with a liquid–liquid phase separator.

In the case of n-propyl propionate synthesis, Altman et al. (2010) already performed RD experiments in a pilot-scale column. However, the authors noted several operational difficulties, especially in the context of the liquid–liquid phase separator employed. A data reconciliation test to identify the stationary operating conditions confirmed these difficulties. They reported that nearly 40% of their performed experiments failed the data-reconciliation test. The reconciliation test is passed if all the measured values can be adjusted within the limits of the experimental error in such a way that physical constraints are satisfied (i.e., mass and component balances and reaction rates). Based on these results, we decided to install a new liquid–liquid phase separator in the pilot plant (Altman et al., 2010) to improve the phase separation and to provide more reliable experimental results for *n*-propyl propionate synthesis. Especially, the crucial need for experimentally determined column profiles has recently been emphasised by Taylor (2006). The purpose of our article is to fill this gap and to provide reliable experimental temperature and concentration profiles for *n*-propyl propionate synthesis in a reactive distillation column coupled with a liquid–liquid phase separator. Furthermore, a nonequilibrium-stage model is used to predict the experimental results. The comparison between experimental and simulated column profiles is discussed in detail.

REACTION SYSTEM

The chemical system investigated in this study is the heterogeneously catalysed synthesis of *n*-propyl propionate (ProPro). It is formed by the reversible, acid-catalysed, liquid-phase esterification of 1-propanol (POH) and propionic acid (ProAc) with water as an additional by-product according to Eq. (1):

$$\tag{1}$$

Based on the work of Buchaly et al. (2007), the strongly acidic ion-exchange resin Amberlyst 46™ from Rohm and Haas was selected as a heterogeneous acid catalyst for *n*-propyl propionate synthesis. Because Amberlyst 46™ exhibits active centres only at its surface, undesired side reactions, such as the dehydration of 1-propanol, leading to the corresponding alkenes, and the self-condensation of 1-propanol, leading to the corresponding ether, can be suppressed (Duarte et al., 2006).

The nomenclature and the boiling points of the pure components at atmospheric pressure are listed inTable 1. The chemical system shows a complex thermodynamic phase behaviour, which is illustrated by the azeotropic data presented in Table 2. In total, two homogeneous and two heterogeneous azeotropes exist. The low-boiler of the system is the ternary, heterogeneous azeotrope consisting of water, 1-propanol and n-propyl propionate.

Table 1: Nomenclature and pure component boiling points at $p=1$ atm (NIST, 2009)

Component	IUPAC name	Formula	T_b (K)
POH	1-propanol	C_3H_7OH	370.4
Water	Water	H_2O	373.2
ProPro	Propanoic acid propyl ester	$C_2H_5COOC_3H_7$	396.1
ProAc	Propanoic acid	C_2H_5COOH	414.1

Table 2: Azeotropic data at $p=1$ atm (Gmehling et al., 2004)

Type	Molar fraction (mol/mol)				T_b (K)
	x_{POH}	x_{water}	x_{ProPro}	x_{ProAc}	
Heterogeneous	0.350	0.520	0.130	–	323.4
Homogeneous	0.431	0.569	–	–	390.9
Heterogeneous	–	0.650	0.350	–	363.2
Homogeneous	–	0.950	–	0.050	373.1

A large miscibility gap exists in the subsystem consisting of water, 1-propanol and n-propyl propionate. The relevant ternary phase diagram at atmospheric pressure is shown in Fig. 1. Additionally, the distillate compositions of all the experiments performed herein (Exp1–Exp7) are illustrated in Fig. 1. Because the miscibility gap increases with lower temperatures, the distillate stream was cooled down to a temperature of around 293 K before entering the liquid–liquid separator to enhance the phase separation. Therefore, the binodal curve in the ternary phase diagram was calculated at that temperature.

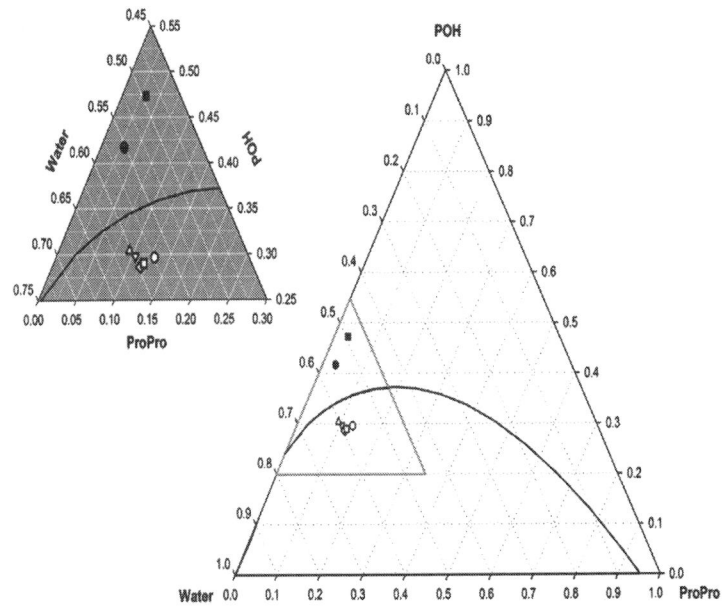

Figure 1: Ternary phase diagram for the subsystem comprising water, 1-propanol (POH) and *n*-propyl propionate (ProPro). The miscibility gap is bound by a binodal curve (—), which was calculated at a temperature of 293 K. The distillate stream compositions of all the experiments are indicated as follows: ■ Exp1, ● Exp2, ▼ Exp3, □ Exp4, ○ Exp5, ▲ Exp6 and ◊ Exp7.

EXPERIMENTAL

Column Setup

The experiments were performed in a pilot-scale reactive distillation column. Fig. 2 shows the experimental setup. The column has an inner diameter of 50 mm, is made of glass and consists of six segments. The adiabatic process conditions in the column are maintained by the insulation with two layers of mineral wool.

Additionally, a heating wire is installed between the layers along each glass segment and is set to the average vapour temperature, which is calculated from the online measured temperatures above and below each segment. For evaporation, a naturally circulating reboiler with a 2-L holdup and a maximum electrical heat duty of 6.6 kW is installed.

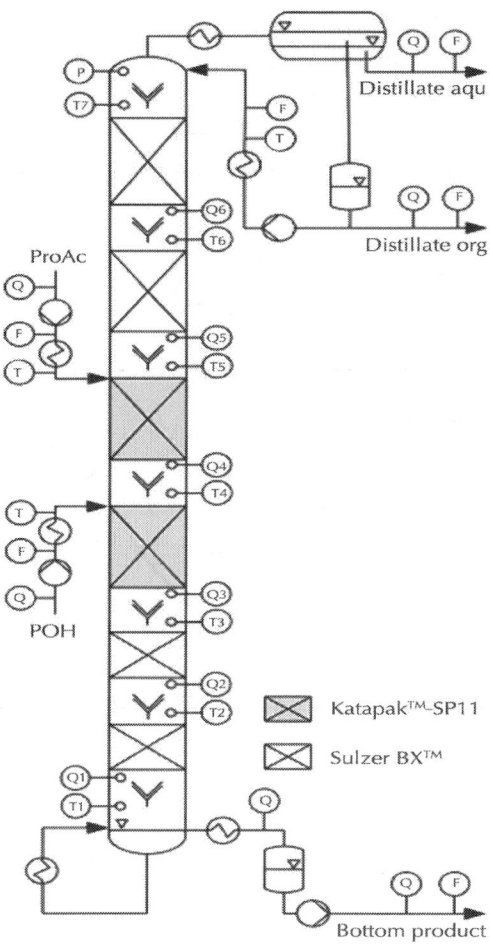

Figure 2: Experimental setup.

Two structured packing materials are incorporated into the column. The reactive section consists of two segments of the structured catalytic packing Katapak™-SP 11, each of 1.05 m in height. In the catalytic packing, the selected heterogeneous catalyst Amberlyst™ 46 is immobilised. The mass of the dry catalyst per metre of catalytic packing ($_{mcat,dry}$=0.205 kg/m) was experimentally determined by Buchaly et al. (2007). Conventional structured Sulzer BX™ packing elements are used in the stripping section of 1 m and in the rectifying section of 2.4 m.

The high-boiling component, propionic acid, was fed above the reactive section, and 1-propanol was fed 1.05 m below the acid feed to establish a countercurrent flow and extensive contact between the reactants. Because the product, n-propyl propionate, is the second high-boiling component in the quaternary system, its purity is limited by the conversion of propionic acid, the high-boiling component in the system. According to Le Chatelier's principle, the alcohol was fed in molar excess to increase the conversion of propionic acid. The relevant characteristics of the pilot-scale column are listed in Table 3.

Table 3: Characteristics of pilot-scale reactive distillation column coupled with a liquid–liquid phase separator

Column diameter	50 mm
1-propanol feed	2.1 m
Propionic acid feed	3.1 m
Stripping section	1 m (Sulzer BX™)
Reactive section	2.1 m (Katapak™ SP 11)
Rectifying section	2.4 m (Sulzer BX™)
Mass of dry catalyst per metre of catalytic packing	0.205 kg/m
Condenser type	Total
Reboiler type	Naturally circulating reboiler
Operating pressure	Atmospheric
Liquid–liquid phase separator operated at	293 K

Because propionic acid is the high-boiling component, only negligible traces occurred at the top of the column. Therefore, the distillate stream consisted of 1-propanol, water and *n*-propyl propionate. The relevant ternary phase diagram at atmospheric pressure shows a miscibility gap (see Fig. 1). This miscibility gap in the ternary system was used to recover unreacted 1-propanol from the distillate stream by liquid–liquid phase separation (described in the following Section 3.2). Therefore, suitable operating conditions had to be selected to ensure that the condensate composition at the top of the column fell into the region in which phase splitting occurs. To enhance this phase splitting, the distillate stream was cooled down to a temperature of 293±3 K before entering the liquid–liquid separator. While the aqueous phase was completely withdrawn and a small fraction of the organic phase was purged, the majority of the organic phase, containing unconverted 1-propanol, was returned to the column as reflux.

To store and monitor all the process variables and to control all the operating variables during the experiments, a process control system (Simatic™ PCS7, Siemens AG) was installed. The mass-flow rates of the feeds, the bottom and the aqueous distillate were determined with industrial balances via the time derivative of the mass change. Additionally, calibrated Coriolis-type flow metres were installed to measure the organic distillate and the reflux flow rate. A pressure transducer recorded the top pressure of the reactive distillation column. As described above, the reactive distillation column consists of six packing sections. A liquid distributor is installed above each packing section to ensure good distribution. At each distributor, temperatures in the vapour phase were measured by calibrated PT-100 thermocouples. Additionally, the temperatures of the vapour flow entering the condenser and that above the reboiler, the temperature in the liquid–liquid phase separator and the temperature of the reflux stream entering the reactive distillation column were recorded. Liquid samples were withdrawn with a syringe at each distributor as well as from the bottom stream and both distillate streams (organic and aqueous). Table 4 summarises all the temperature and sample locations along the column.

Table 4: Temperature and sample locations along the column

Location from column bottom (m)	Temperature (vapour)	Sample (liquid)
0	T1	Q_1
0.5	T2	Q_2
1	T3	Q_3
2.05	T4	Q_4
3.1	T5	Q_5
4.3	T6	Q_6
5.5	T7	Q_7

Liquid–Liquid Phase Separator

To continuously separate the distillate stream leaving the reactive distillation column into aqueous and organic phases, a horizontal liquid–liquid phase separator was installed (DN50, Type A, supplied by QVF Miniplant). The separator has an internal liquid holdup of 0.8 L and is made of glass. An internal overflow valve for the heavy aqueous phase controls the interphase, while an integrated jacket tube holds the light organic phase back from the height-adjustable overflow opening. Therefore, the adjustment of a desired phase ratio and the variation of the residence time for each phase are possible. As mentioned in Section 3.1, the distillate stream was cooled down to a temperature of 293±3 K before entering the liquid–liquid separator to enhance the phase splitting. A PT 100 thermocouple is used to measure the temperature inside the separator. A sufficiently high residence time for both the aqueous and organic phases was necessary to achieve a complete phase split in the liquid–liquid separator. Preliminary tests showed that a residence time of at least 10 min for both phases was required.

Experimental Procedure

In each experiment, a stable steady-state operating point of the reactive distillation column was reached. To shorten the necessary start-up time to reach this operating point, a defined start-up procedure was followed, as described by Altman et al. (2010). After the start-up procedure, the column and the liquid–liquid separator transitioned into the desired steady-state operating point. In our experiments, a steady-state operating point was defined by the following requirements:

- The temperature at each measured location is constant within $T=\pm1$ K.
- The overall mass and component balances as well as reaction rates in the RD column equipped with a liquid–liquid separator must be fulfilled.

This steady-state condition was checked by applying data reconciliation, which is explained in Section 3.5. After reaching and identifying the steady-state operating point, liquid samples from the six liquid distributors of the column were taken. These samples and the samples of the bottom stream and both distillate streams (organic and aqueous) provided detailed data on the concentration profile. In total, three profiles were taken at 60-min intervals during every experiment.

Analytical Methods and Chemicals

A gas chromatograph supplied by Shimadzu (GC-14B) was used to analyse all the samples offline. The instrument was equipped with a flame-ionisation detector (FID), and helium was used as a carrier gas with a gas velocity of 27 cm/s. An FS-Innopeg 1000 capillary column (25 m×25 m) with a film thickness of 0.27 μm was used. The split ratio was set to 1:4.5 min, and an optimised temperature programme was implemented (333 K for 4.5 min, a ramp of 30 K/min to 453 K followed by a hold for 2.4 min); typical retention times were 3.6 min for 1-propanol, 3.8 min for n-propyl propionate and 9.1 min for propionic acid. Because the FID was only able to analyse the mass fraction of the organic components, Karl Fischer titration (Mettler Toledo DL 31) was used to determine the water content.

The chemicals, 1-propanol (Syskem Chemie Ltd., Wuppertal, Germany) and propionic acid (Silbermann Ltd., Gablingen, Germany), were delivered in industrial bins. The purities of the chemicals in the feed bins of the reactive distillation column were determined prior to each experiment. On average, the purities of 1-propanol and propionic acid were higher than 99.6 mass percent.

Data Reconciliation

The measured process values are inevitably corrupted by errors during the measurement itself and during the processing and transmission of the measured signal to the process control system. The closing of the process mass and component balances can be quite difficult due to these errors. Hence, data reconciliation was performed to adjust the measured values considering their experimental standard errors such that they satisfied the constraints of the mass and component balances and reaction rates (Narasimhan and Jordache, 1999). The so-called reconciled values thus represent the statistically most probable process state and allow for a unique and consistent description of the system. In general, data reconciliation can be formulated by a constrained weighted least-squares optimisation problem, as previously applied to the reactive distillation systems by Buchaly (2009). Here, we report the application of data reconciliation to a reactive distillation column coupled with a liquid–liquid phase separator. The objective function defines the totally weighted sum square of the adjustments made to all the nM measurements and is formulated in Eq. (2), in which $z_{i,exp}$ is the experimental value, $z_{i,rec}$ is the reconciled value and z is the experimental standard deviation for each considered measured variable z_i. The standard deviation corresponds to the uncertainty of the measured variable and must be experimentally determined or specified to a reasonable value

$$\Phi = \min \sum_{i=1}^{n_M} \frac{(z_{i,exp} - z_{i,rec})^2}{\sigma_{z_i}^2}$$

(2)

$$\frac{\partial m}{\partial t} = \dot{m}_{F,ProAc} + \dot{m}_{F,POH} - (\dot{m}_{D,aqu} + \dot{m}_{D,org}) - \dot{m}_B = 0$$

(3)

$$\sum_{i=1}^{n_C} w_{F,ProAc,i} = 1; \quad \sum_{i=1}^{n_C} w_{F,POH,i} = 1; \quad \sum_{i=1}^{n_C} w_{B,i} = 1; \quad \sum_{i=1}^{n_C} w_{D,aqu,i} = 1;$$

(4)

$$\sum_{i=1}^{n_C} w_{D,org,i} = 1$$

(3)

$$\frac{\partial n_i}{\partial t} = \dot{n}_{i,F,ProAc} + \dot{n}_{i,F,POH} - (\dot{n}_{i,D,aqu} + \dot{n}_{i,D,org}) - \dot{n}_{i,B} = r \quad i = 1 \cdots n_C$$

(5)

Sartorius MC1 industrial balances were used to measure both feed-flow rates, $\dot{m}_{F,ProcAc}$ and $\dot{m}_{F,POH,}$, the aqueous distillate flow rate $\dot{m}_{D,aqu}$ and the bottom flow rate $\dot{m}_B$. For these accurate balances, the standard deviation of the measurement with a value of ±0.01 kg/h was used for the data reconciliation. A Coriolis-type flow metre from Rhenonik (RHE 08), equipped with a Rheonik RHM 015 sensor, determined the reflux flow rate $\dot{m}_R$ and the organic distillate flow rate $\dot{m}_{D,arg}$. An absolute error of ±0.02 kg/h was specified for the data reconciliation.

Eqs. (3), (4) and (5) define the set of model constraints to be satisfied by the reconciled measured values. If the reactive distillation column combined with the liquid–liquid phase separator is at a steady state, the overall mass balance must be zero. Additionally, the summation condition of the mass fractions must be fulfilled for all the streams. Furthermore, the reaction rates of all the components must be equal due to the stoichiometry of the n-propyl propionate synthesis. Table 5 summarises the error of the measurements considered in this work. The data reconciliation problem was implemented in Microsoft Excel™ and was solved using the Excel Solver, which applies a *generalised reduced gradient* algorithm (Lasdon et al., 1978).

Table 5: Uncertainties of the measurements used for data reconciliation

Mass fraction $_{wi}$ ≤0.1 g/g	±0.005 g/g
Mass fraction $_{wi}$ >0.1 g/g	±0.01 g/g
Flow rate from balance	±0.01 kg/h
Flow rate from Coriolis-type flow metre	±0.02 kg/h

Two criteria were used to determine when the data reconciliation test is passed and thus the measured values can be regarded as reconciled. First, all the measured values may only be adjusted within the limits of the experimental error:

$$(z_{i,exp} - z_{i,rec})^2 \leq \sigma_{z_i}^2 \quad i = 1 \cdots n_M$$

$$(6)$$

Second, according to the VDI guideline 2048 of the Association of German Engineers (The Association of German Engineers, 2000) the calculated objective function Φ must be lower than the 95% quantile of the χ^2-distribution from the degree of freedom f of the formulated optimisation problem:

$$\Phi \leq \chi_{f,95\%}^2$$

$$(7)$$

The degree of freedom f is the difference between the number of measured values and of equations used in the data reconciliation test. The 95% quantile of the χ^2-distribution can be found in (The Association of German Engineers, 2000) or in statistical tables. In the case of the reconciliation test applied for n-propyl propionate synthesis, a degree of freedom with a value of 15 (25 measured values, 10 equations) can be calculated. According to the χ^2-distribution, the objective function Φ must be lower than 24.95. If one of these two criteria, formulated in Eqs. (6) and (7), is not satisfied, the measured data were rejected and not used for further model validation.

MODELLING AND SIMULATION

The modelling of heterogeneously catalysed reactive distillation processes has been comprehensively reviewed by several authors (Taylor and Krishna, 2000 and Sundmacher and Kienle, 2003). The mathematical models describing the reactive distillation processes can be primarily classified as either equilibrium-stage (EQ) or nonequilibrium-stage (NEQ) models. The EQ model assumes that both bulk phases are perfectly mixed and that the streams leaving the stage are in thermodynamic equilibrium. In contrast, the NEQ model takes into account the actual rates of multi-component mass and heat transfer as well as the process hydrodynamics (Baur et al., 2000).

In this work, a NEQ model (Taylor and Krishna, 2000, Górak and Hoffmann, 2001 and Kloeker et al., 2005) was applied, which has been validated with experimental results for n-propyl propionate synthesis in a reactive distillation column by Buchaly et al. (2007). The mass and heat transfer across the vapour–liquid interface was described using the two-film theory (Lewis and Whitman, 1924). The multi-component diffusion in the films was modelled by the effective-diffusivity approach (Hougen, 1947), and the Chilton–Colburn analogy (Chilton and Colburn, 1935) was applied for the calculation of the heat transfer rates. The hydrodynamics of the column internals were considered by the application of packing specific correlations (Rocha et al., 1993, Bravo et al., 1985 and Brunazzi and Viva, 2006). The heterogeneously catalysed reaction was described by a pseudohomogeneous approach assuming no mass transfer resistance between the liquid phase and catalyst. The liquid–liquid phase separator was modelled as an ideal, adiabatic phase separator applying a three-phase flash calculation neglecting the vapour phase at fixed temperature and pressure. The UNIQUAC model was applied to consider the nonideal liquid-phase behaviour (see next section). All the model equations were implemented in the commercial equation-oriented simulation environment Aspen Custom Modeler™.

Thermodynamic and Physical Properties

The applied simulation environment Aspen Custom Modeler™ uses an interface with Aspen Plus™ for the calculation of thermodynamic and physical properties (see Table A.1). The UNIQUAC model (Abrams and Prausnitz, 1975) was employed for the vapour–liquid equilibrium (VLE) calculations in Aspen Plus™. The Hayden O'Connel equation of state (Hayden and O'Connell, 1975) was used to account for the self-association of propionic acid in the vapour phase. The complete set of binary interaction parameters for the VLE calculations was previously presented by Buchaly et al. (2007) and is listed in Table A.2 of Appendix A. For the liquid–liquid equilibrium (LLE) description in the phase separator, a second set of UNIQUAC interaction parameters available in Aspen Plus™ was used. The LLE calculation was checked against experimental data extracted from the Dortmund Data Bank (Sørensen and Arlt, 1980). The complete set of applied binary interaction parameters for the LLE calculations with the UNIQUAC model is given in the publication by Altman et al. (2010) and is additionally summarised in Table A.3 of Appendix A.

Reaction Kinetics

Duarte et al. (2006) performed experiments in the expected temperature range of the column operation (363–383 K) to determine the reaction equilibrium and kinetics for the applied heterogeneous catalyst Amberlyst 46™. To describe the reaction mathematically, a pseudohomogeneous approach formulated in activities, ai, for the reaction rate of component i was used, as given by

$$r_i = \frac{dn_i}{dt} = v_i m_{cat,dry} C_{act} \left(k_1(T) a_{ProAc} a_{POH} - \frac{k_1(T)}{K_a(T)} a_{ProPro} a_{water} \right)$$

(8)

$$k_1(T) = 7.381 \times 10^7 \exp\left\{ -\frac{5.963 \times 10^4}{R_m T} \right\}$$

(9)

$$K_a(T) = 6.263 \exp\left\{\frac{4.519 \times 10^3}{R_m T}\right\}$$

(10)

The kinetic expression includes the dry catalyst mass $_{mcat,dry}$ and the concentration of active sites $_{cact}$. Both values were experimentally determined in a prior work (Buchaly et al., 2007) as $_{mcat,dry}$=0.205 kg/m and $_{cact}$=0.78 eq/kg. The temperature dependency of the equilibrium constant $_{Ka}(T)$ and the rate constant $_{k1}(T)$ are taken into account according to the Arrhenius equation. The UNIQUAC model to calculate the liquid phase activity coefficients ai as described in Section 4.1 is also used for the determination of the chemical equilibrium and the reaction kinetics.

To provide an overview of the applied nonequilibrium-stage model, important information on the used model and its model parameters are given in Appendix A (see Table A.4).

RESULTS AND DISCUSSION

Overview of the Experiments

To provide reliable experimental data for the model validation, seven experiments (Exp1–Exp7) on the synthesis of n-propyl propionate synthesis were successfully performed in a pilot-scale reactive distillation column. In Exp1 and Exp2, a single phase was produced in the liquid–liquid phase separator, and the results are discussed in detail in Section 5.2. The objective of Exp1 and Exp2 was to verify the activity of the immobilised heterogeneous catalyst Amberlyst 46™. The catalyst was immobilised five years ago byBuchaly (2009) and has been applied in numerous experimental studies on the synthesis of n-propyl propionate (approximately 800 operating hours). Because Buchaly (2009) carried out experiments in the absence of a subsequent liquid–liquid phase separator, we decided to first check the catalyst activity by conducting two experiments (Exp1 and Exp2) without an occurring phase split in the liquid–liquid phase separator. Alternatively, Exp3–Exp7 were carried out such that a phase separation of the distillate stream was achieved. These experiments are presented in Section 5.3.

An overview of the operating conditions and molar-flow rates applied in the experiments is provided inTable 6. In all of the experiments, the stoichiometric ratio of 1-propanol to propionic acid was set to 1.5 to increase the conversion of propionic acid, the high-boiling component. The molar reflux ratio RR was varied from 1.65 to 2.51, and the molar distillate-to-feed ratio D/F was adjusted between 0.37 and 0.59. Exp1–Exp6 were carried out at a total feed flow rate of 61.2 mol/h, whereas Exp7 was performed at a total feed-flow rate of 53.8 mol/h to investigate the column behaviour at low gas and liquid loads. Phase separation was not observed in Exp1 and Exp2 (see Table 6), because the operating conditions of the reactive distillation column were selected such that the distillate stream leaving the column had a composition outside of the miscibility gap.

This fact is illustrated in Fig. 1, in which the molar compositions of the distillate streams for Exp1–Exp7 are indicated in the ternary phase diagram. Only a narrow operational range was covered by our experiments because the investigated operational parameters were determined under three boundary conditions: liquid–liquid phase separation was prohibited inside the column, liquid–liquid phase separation was promoted in the separator and the reactive section of the column was maintained at temperatures less than 393 K, in line with the maximum operating temperature of the catalyst.

Table 6: Operating conditions and molar-flow rates of all the experiments carried out in the pilot-scale reactive distillation column. Exp1 and Exp2 were performed without a phase split in the liquid–liquid separator

	POH-feed (mol/h)	ProAc-feed (mol/h)	Reflux ratio RR(mol/mol)	Distillate-to-feed ratio D/F(mol/mol)	Distillate-organic (mol/h)	Distillate-aqueous (mol/h)	Bottom (mol/h)	Temperature in separator (K)	Status data reconciliation
Exp1[a]	36.672	24.510	2.51	0.588	35.950	–	25.232	–	Correct
Exp2[a]	36.822	24.508	2.51	0.532	32.598	–	28.732	–	Correct
Exp3	36.736	24.568	1.68	0.406	10.464	14.433	36.406	293.8	Correct
Exp4	36.657	24.526	2.18	0.432	16.821	9.628	34.735	292.3	Correct
Exp5	36.687	24.521	2.35	0.368	6.046	16.469	38.693	291.9	Correct
Exp6	36.723	24.575	2.08	0.467	21.766	6.891	32.641	293.3	Correct
Exp7	31.824	21.930	1.65	0.463	12.081	12.809	28.864	290.7	Correct

[a]Without subsequent liquid–liquid phase split.

Side-products such as di-*n*-propyl ether were not detected in any of the pilot-scale experiments. Therefore, in the *n*-propyl propionate synthesis by reactive distillation, the formation of side-products can be neglected when the heterogeneous catalyst Amberlyst 46™ is used under the proposed conditions. Similarly, Buchaly et al. (2007) and Altman et al. (2010) did not observe side-product formation in previous experimental studies.

Reactive Distillation Column without Coupled Liquid–Liquid Phase Separation

Results of Data Reconciliation

Exp1 and Exp2 passed the data reconciliation test and, hence, showed high accuracy. The raw experimental data and reconciled experimental data are compared in Table 7 for the example of Exp1. During the data reconciliation process, only minor modifications to the raw experimental data were necessary. The experimentally determined compositions and temperature profiles of the column for Exp1–Exp2 are available online as supplementary material to the electronic version of this article.

Table 7: Comparison between raw and reconciled experimental data for Exp1

	POH-feed		ProAc-feed		Distillate		Bottom	
	Raw data	Rec. data	Raw data	Rec. data	Raw data	Rec. data	Raw data	Rec. data
$\dot{n}$ (mol/h)	36.689	36.672	24.528	24.510	35.818	35.950	25.367	25.232
x_{POH} (mol/mol)	0.997	1.000	0.000	0.000	0.479	0.473	0.089	0.085
x_{water} (mol/mol)	0.003	0.000	0.012	0.012	0.489	0.495	0.000	0.001
x_{ProPro} (mol/mol)	0.000	0.000	0.000	0.000	0.032	0.032	0.651	0.649
x_{ProAc} (mol/mol)	0.000	0.000	0.988	0.988	0.000	0.000	0.260	0.265

Model Validation

The theoretical analysis of a reactive distillation process is a challenging task because it requires a detailed description of multicomponent mass transfer effects, vapour–liquid equilibria and heterogeneously catalysed reaction. As described in Section 4, a nonequilibrium-stage model was applied in the present study. It must be emphasised that the simulations with the nonequilibrium-stage model are fully predictive.Fig. 3 compares the experimental and simulated composition and temperature profiles for the example of Exp2. Prior to further discussion of these results, it should be noted that accurate sampling was not possible with the propionic acid feed distributor due to configurational issues. Therefore, the results obtained from the sampling point at a packing height of $h=3.1$ m are not shown in Fig. 3. The symbols represent the experimental values, and the lines represent the simulated values of the nonequilibrium-stage model. Excellent agreement between the experimental and simulated composition and temperature profiles was observed along the entire packing height, which illustrates the suitability of the nonequilibrium-stage model at describing the reactive distillation process. These results agree with those presented byBuchaly (2009) and demonstrate that the catalyst Amberlyst 46™ does not show a considerable loss of activity following its immobilisation in the column.

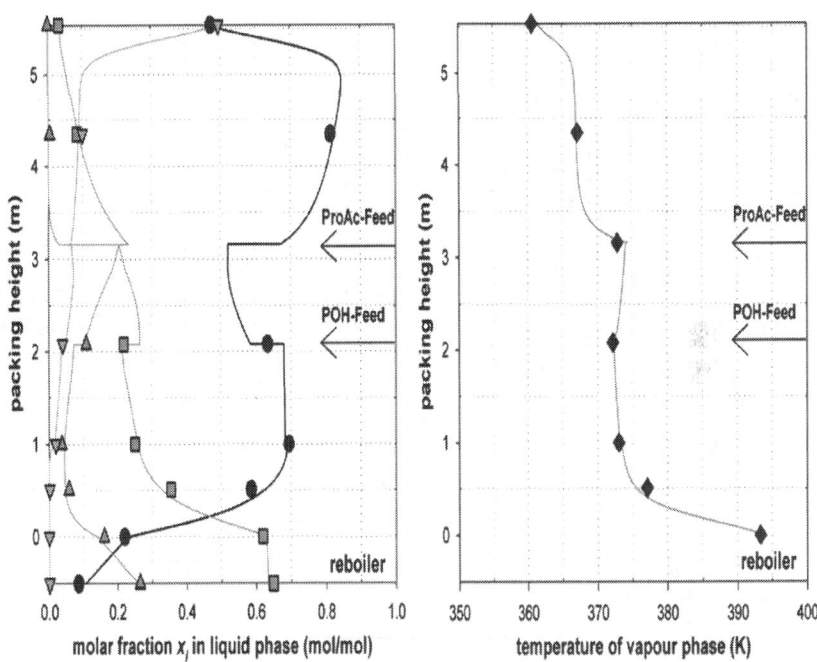

Figure 3: Comparison between the experimental and simulated composition and temperature profiles obtained in Exp2. The symbols represent the experimental values, and lines represent the simulated results. Left: molar fraction in the liquid phase: -●- 1-propanol, -▼- water, -■- *n*-propyl propionate and -▲- propionic acid. Right: temperature profiles of the vapour phase.

Reactive Distillation Column with Coupled Liquid–Liquid Phase Separation

Results of Data Reconciliation

Exp3–Exp7 were carried out such that the distillate stream leaving the column had a composition within the miscibility gap. In this case, liquid–liquid phase separation was achieved. Reliable experimental data were obtained because all of the experiments passed the data-reconciliation test. In particular, the horizontal liquid–liquid phase separator used to separate the phases contributed to the reliability of the experimental results. Altman et al. (2010) investigated the synthesis of n-propyl propionate in the same reactive distillation column but employed a vertical liquid–liquid phase separator. A comparatively poor performance with respect to data reconciliation was reported. Namely, nearly 40% of the experiments failed the data-reconciliation test and thus could not be used for model validation. In contrast, the horizontal separator used in the present study allowed for the adjustment of the residence time. Thus, the residence time was increased to a mean value of 10 min, and improved phase separation was observed. Accordingly, highly accurate experimental results for the synthesis of n-propyl propionate in a pilot-scale reactive distillation column coupled with a liquid–liquid phase separator were generated. The raw experimental data and reconciled experimental data are exemplarily shown for Exp7 in Table 8. The experimentally determined composition and temperature profiles of the column for Exp3–Exp7 are available online as supplementary material to the electronic version of this article.

Table 8: Comparison between raw and reconciled experimental data for Exp7

	POH-feed		ProAc-feed		Distillate-organic		Distillate-aqueous		Bottom	
	Raw data	Rec. data	Raw data	Rec. data	Raw data	Rec. data	Raw data	Rec. data	Raw data	Rec. data
$\dot{n}$ (mol/h)	31.686	31.824	21.803	21.930	12.205	12.081	13.059	12.809	29.216	28.864
x_{POH} (mol/mol)	0.997	0.991	0.000	0.000	0.351	0.353	0.059	0.059	0.327	0.324
x_{water} (mol/mol)	0.001	0.009	0.012	0.020	0.490	0.485	0.939	0.939	0.000	0.000
x_{ProPro} (mol/mol)	0.000	0.000	0.000	0.000	0.160	0.161	0.002	0.002	0.521	0.527
x_{ProAc} (mol/mol)	0.000	0.000	0.988	0.980	0.000	0.000	0.000	0.000	0.151	0.150

Model Validation

In the present study, the synthesis of *n*-propyl propionate was successfully performed in a pilot-scale reactive distillation column coupled with a liquid–liquid phase separator. As presented in Section 4, a nonequilibrium-stage model was applied to simulate the reactive distillation column, and a three-phase flash calculation was used to simulate liquid–liquid phase separation. Fig. 4 compares the experimental and simulated composition and the temperature profiles of the reactive distillation column for the example of Exp3. The symbols represent experimentally determined values, and the lines represent the results of the nonequilibrium-stage model. Satisfactory agreement between the experimental and simulated composition and temperature profiles was obtained for Exp3. However, at a packing height of 4.3 m, an unusually large deviation between the experimental and simulated values was observed. Specifically, the molar fraction of 1-propanol and water was not correctly predicted by the applied model, and relative deviations of approximately 30% were obtained. Disagreement between the experimental data and the results of the model was consistently observed when phase separation occurred in the liquid–liquid separator (i.e., Exp3–Exp7). On the contrary, deviations between the experimental and simulated results were not observed for Exp1 and Exp2, where liquid–liquid phase separation did not occur in the separator (see Section 5.2.2). Apparently, the proposed model cannot predict the composition and temperature profile in the rectifying section of the column when phase separation occurs in the liquid–liquid separator. Poor results may be obtained under these conditions because the three-phase flash calculation assumes that phase separation is ideal, which is not observed experimentally. Therefore, the simulation of the phase split in the liquid–liquid phase separator is analysed hereafter.

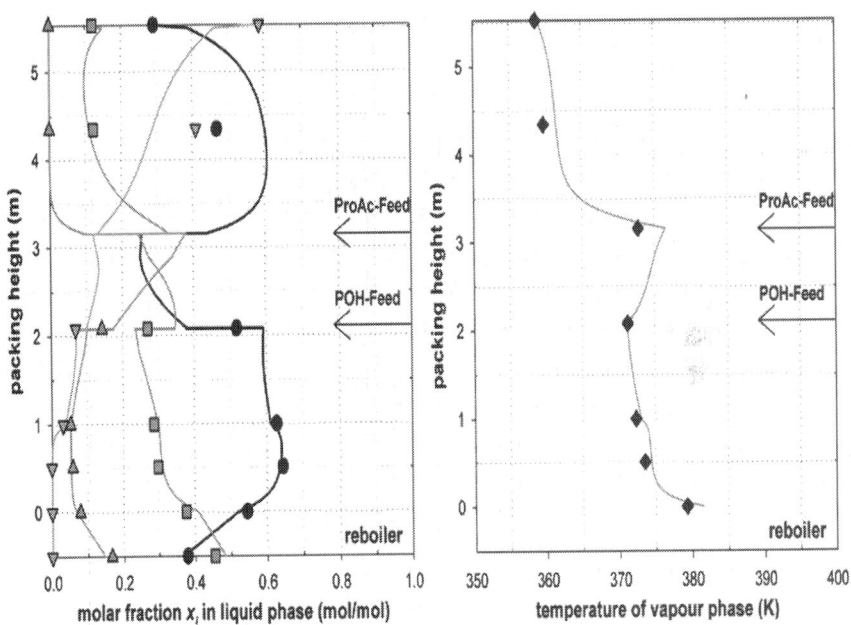

Figure 4: Comparison between the experimental and simulated column profiles obtained in Exp3. The symbols represent experimental values, and lines represent the simulated results. Left: molar fraction in the liquid phase: -●- 1-propanol, -▼- water, -■- n-propyl propionate and -▲- propionic acid. Right: temperature profiles of the vapour phase.

The experimentally determined and simulated compositions of the distillate stream and both phases leaving the separator in Exp3 are shown in the ternary diagram in Fig. 5. Only minor deviations between the experimental and simulated results were observed. The experimental compositions of both phases were very similar to the compositions obtained from the theoretical binodal curve. Minor deviations between the experimental and theoretical values may be attributed to the fact that the experimental values are subject to measurement errors. Moreover, the binodal curve was calculated with a thermodynamic model, which may lead to inaccurate results. Nevertheless, Fig. 5 illustrates that a nearly ideal phase separation occurred in the liquid–liquid phase separator in Exp3. Therefore, the application of a three-phase flash calculation should actually be justified to simulate the liquid–liquid phase split. To verify that complete phase separation occurred, a portion of the organic phase leaving the separator was collected and was allowed to stand overnight. The formation of an aqueous phase was not observed. However, the reader must keep in mind that during the experiment the majority of the organic phase is returned to the reactive distillation column as a reflux. As shown in Table 9, minor differences between the experimentally determined and simulated reflux-stream compositions were observed when the three-phase flash calculation was applied. Hence, the question arises whether such slight deviations between the experimental and simulated reflux-stream compositions may lead to significantly different internal column profiles and are thus responsible for the deviation between simulated and experimental column profiles.

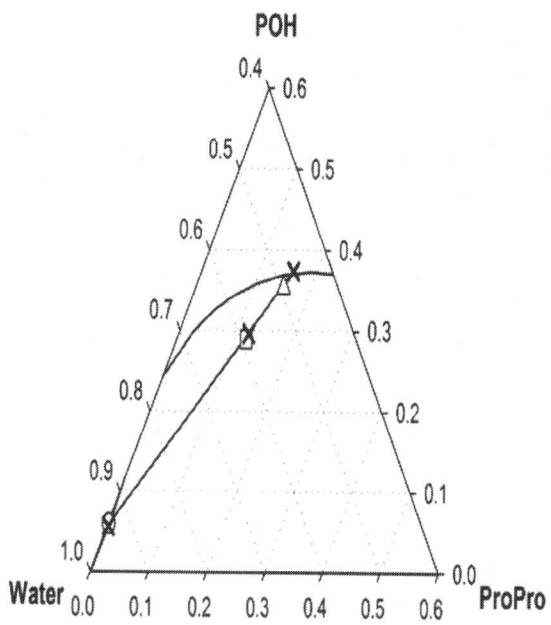

Figure 5: Comparison between the experimental and simulated liquid–liquid phase-separation results for Exp3. The composition of the distillate stream (□) entering the separator and the composition of organic (▲) and aqueous phases (○) leaving the separator are marked. The corresponding

simulated values are labelled with crosses (×) and are connected by the tie line.

Table 9: Experimental and simulated reflux-stream compositions for Exp3

	Reflux-stream composition	
	Experiment	Simulation[a]
x_{POH} (mol/mol)	0.353	0.373
x_{water} (mol/mol)	0.495	0.471
x_{ProPro} (mol/mol)	0.152	0.156
x_{ProAc} (mol/mol)	0	0

[a]Liquid–liquid phase separation is predicted with three-phase flash calculation.

To determine whether slight deviations between the experimental and simulated reflux-stream compositions lead to significantly different internal column profiles, a modified model was applied. In the proposed model, liquid–liquid phase separation was calculated with component-specific partition ratios instead of a three-phase flash calculation. The definition of the applied partition ratios αi is formulated in Eq.(11), in which $w_{i,aqu,exp}$ is the experimental value of the weight fraction of component i in the aqueous phase and $w_{i,org,exp}$ is the respective weight fraction in the organic phase

$$\alpha_i = \frac{w_{i,aqu,exp}}{w_{i,org,exp}} \quad i = 1 \cdots n_C$$

(11)

Because the partition ratios must be calculated after each experiment, the model loses its predictability. However, predefining partition ratios allows the user to adapt the simulated phase separation in such a way that the experimentally determined composition of the reflux stream entering the reactive distillation column can be reproduced. Therefore, to quantify the effect of the observed deviations between the experimental and simulated reflux-stream composition on the internal column profiles, Exp3 was additionally simulated by applying a model based on partition ratios for the calculation of the liquid–liquid phase separation. It has to be emphasised that the partition ratios were not applied to fit the simulated results to the experimental values. Rather the partition ratios were used to investigate the sensitivity of the internal column profiles to minor changes in the composition of the reflux stream.

The internal column profile of Exp3, which was simulated with the model using partition ratios for the liquid–liquid separation, is shown in Fig. 6. Additionally, the experimental results and the simulated results obtained from the model containing a three-phase flash calculation for the liquid–liquid separation are also provided. As already illustrated in Table 9, there are only minor differences in the reflux-stream compositions predicted with the three-phase flash procedure and calculated with experimentally determined partition ratios. However, these slight differences have a considerable effect on the internal column profiles, as shown in Fig. 6. Significant differences between the results of both models were observed because completely different concentration profiles were calculated in the rectifying section of the column. In addition, the temperature profile is sensitive to the composition of the reflux stream in the upper part of the column. In summary, the simulations verified that minor changes in the composition of the reflux stream can lead to significantly different internal column profiles. Therefore, the observed deviation between the experimentally determined and simulated column profiles, illustrated in Fig. 4, can be attributed to the sensitivity of the operating point.

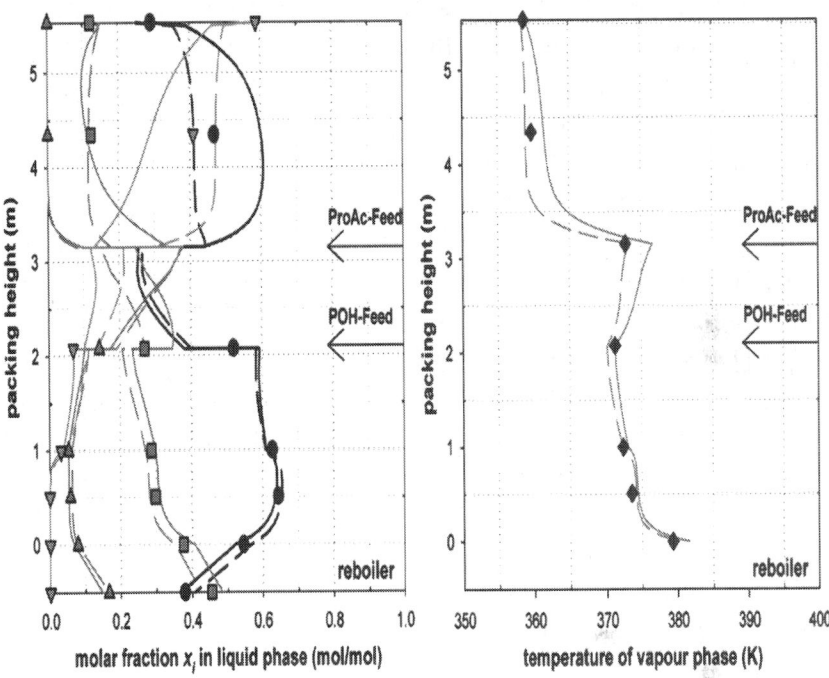

Figure 6: Internal column profiles obtained in Exp3. The column profiles were simulated by applying a model based on experimentally determined partition ratios (dashed lines) and a model containing a three-phase flash calculation (solid lines) for the simulation of phase separation in the liquid–liquid separator. Left: molar fractions in the liquid phase: -●- 1-propanol, -▼- water, -■- n-propyl propionate and -▲- propionic acid. Right: temperature profiles of the vapour phase.

Interestingly, the experimental molar fractions of 1-propanol and water at a packing height of 4.3 m were not reproduced by either model and fell between the simulated values. Also, when experimentally determined partition ratios were used to calculate the phase separation, the concentration profile in the rectifying section of the column could not be accurately reproduced. Because experimental values are subject to measurement errors, the actual partition ratios of the liquid–liquid separation and thus the composition of the reflux stream that is recycled back to the column cannot be accurately determined. Moreover, as previously noted, minor changes in the composition of the reflux stream lead to significantly different internal column profiles (see Fig. 6).

To summarise, the predictive model used in this work, which contains a three-phase flash calculation for the simulation of the phase separation in the liquid–liquid separator, cannot predict the internal column profiles with an excellent agreement in the case of *n*-propyl propionate synthesis. Both the thermodynamic prediction of phase separation and a nonideal liquid–liquid separation can lead to minor differences in the experimental and simulated compositions of the reflux stream. Due to the sensitivity of the internal column profile to minor changes in the composition of the reflux-stream, deviations between the experimental and simulated profiles of the reactive distillation column were observed (see Fig. 4). Moreover, also when a non-predictive model based on experimentally determined partition ratios was used to calculate phase separation in the liquid–liquid phase separator, the internal profiles of the reactive distillation column was not accurately reproduced. Due to measurement errors, an exact calculation of the partition ratios and thus of the reflux-stream composition cannot be obtained. Therefore, the observed deviation between the experimentally determined and simulated column profiles can be attributed to the sensitivity of the operating point. Sensitivity was only observed under operating conditions that lead to liquid–liquid phase separation. Alternatively, when phase separation of the distillate stream did not occur, sensitivity to the reflux-stream composition was not observed. In these experiments, the experimental data were in excellent agreement with the simulation results.

CONCLUSIONS

The synthesis of *n*-propyl propionate was successfully performed in a pilot-scale reactive distillation column coupled with a liquid–liquid phase separator. Because the applied horizontal separator allowed for the adjustment of the residence time, which was increased to a mean value of 10 min, the phase separation could be improved in comparison with former results (Altman et al., 2010). Experiments, first performed without an occurring phase separation in the liquid–liquid separator, confirmed that the heterogeneous catalyst Amberlyst 46™ does not show any considerable activity loss after approximately 800 operating hours. Further experiments with an occurring phase separation in the liquid–liquid separator were carried out. All the experiments passed the data reconciliation test and, therefore, provided a reliable set of composition and temperature profiles along the pilot-scale reactive distillation column for further model validation.

During the model-validation process, we identified that the temperature and composition profiles in the rectifying section of the distillation column are highly sensitive to minor changes in the composition of the reflux stream. Therefore, the used predictive model based on a three-phase flash calculation and the non-predictive model based on experimentally determined partition ratios for the simulation of phase separation in the liquid–liquid separator cannot predict the internal column profile with an excellent agreement. This sensitivity and thus the deviations between the experimental and simulated column profiles were observed only under operating conditions that lead to liquid–liquid phase separation. Alternatively, when phase separation did not occur, the experimental data were in excellent agreement with the simulated results.

Our results clearly show that a detailed process analysis of reactive distillation processes applying validated models is essential to identify such sensitivities. The knowledge of existing sensitivities can prevent, for example, the development of instable control designs. These observations have already been made for heterogeneous azeotropic distillation columns, which are known to be difficult to operate and control (Wu and Chien, 2009, Luyben, 2008 and Widagdo and Seider, 1996).

ACKNOWLEDGMENTS

The financial assistance of A. Jantharasuk was provided by SCG Chemicals Co. Ltd. and the Thailand Research Fund (TRF). C. A. Gónzalez-Rugerio is thankful to the CONACyT-DAAD for awarding a Research Fellowship during the course of this work. The authors gratefully acknowledge the support throughout the duration of the project.

APPENDIX A

The applied simulation environment Aspen Custom Modeler™ uses an interface with Aspen Plus™ for the calculation of thermodynamic and physical properties. Applied models are summarised in Table A.1. All abbreviations are in accordance to Aspen. Detailed information can be found in the handbook of Aspen Plus™.

Table A.1: Used models for the calculation of thermodynamic and physical properties in Aspen Plus™

Property	Model name	Property	Model name
PHIVMX	ESHOC	VVMX	ESRK
GAMMA	GMUQUAC	MULMX	MUL2ANDR
WHNRY	WHENRY	KVMXLP	KV2WMSM
PL	PL0XANT	KVLP	KV0STLP
PHIV	ESHOC0	KLMX	KL2SRVR
VL	VL0RKT	DVMX	DV1DKK
HNRY	HENRY1	DLMX	DL1WCA
VLPM	VL1BROC	SIGLMX	SIG2HSS
DHVMX	ESHOC	DGV	ESHOC0
DHVL	DHVLWTSN	DSV	ESHOC0
DHV	ESHOC0	VV	ESHOC0
DGVMX	ESHOC	DL	DL0WCA
DSVMX	ESHOC	DV	DV0DKK
VVMX	ESHOC	MUL	MUL0ANDR

VLMX	VL2IDL	KVPC	KV0STPC
MUVMXLP	MUV2WILK	VV	ESRK0
MUVLP	MUV0CEB	KL	KL0SR
MUVMXPC	MUV2DSPC	SIGL	SIG0HSS

To take into account non-idealities of the liquid phase, the UNIQUAC model was employed. Used binary interaction parameters a_{ij} and b_{ij} for the vapour–liquid equilibrium (VLE) calculations are listed in Table A.2.

Table A.2: UNIQUAC binary interaction parameters a_{ij} and b_{ij} used for VLE calculations (Buchaly et al., 2007)

Component 1	Component 2	i	j	aij	bij (K)
1-Propanol	Propionic acid	1	2	0.0	−145.7
		2	1	0.0	183.9
1-Propanol	Water	1	2	1.8	−668.9
		2	1	−2.4	620.8
1-Propanol	n-Propyl propionate	1	2	0.0	−1.9
		2	1	0.0	−87.3
Water	Propionic acid	1	2	0.0	−244.8
		2	1	0.0	73.8
Water	n-Propyl propionate	1	2	2.2	−887.6
		2	1	6.7	−3266.1
n-Propyl propionate	Propionic acid	1	2	0.0	−229.4
		2	1	0.0	119.9

For the liquid–liquid equilibrium (LLE) description in the ternary subsystem consisting of water, 1-propanol and n- propyl propionate, a second set of UNIQUAC interaction parameters was used. The binary interaction parameters a_{ij} and b_{ij} for the LLE calculations are given in Table A.3.

Table A.3: UNIQUAC binary interaction parameters *aij* and *bij* used for LLE calculations (Altman et al., 2010)

Component 1	Component 2	*i*	*j*	*aij*	*bij* (K)
1-Propanol	Water	1	2	0.0	124.4
		2	1	0.0	−310.4
1-Propanol	*n*-Propyl propionate	1	2	0.0	131.3
		2	1	0.0	−212.5
Water	*n*-Propyl propionate	1	2	1.5	−587.8
		2	1	1.3	−928.2

For the modelling of heterogeneously catalysed *n*-propyl propionate synthesis in a reactive distillation column coupled with a liquid–liquid separator, a nonequilibrium-stage (NEQ) model implemented in the simulation environment Aspen Custom Modeler™ was applied. Table A.4 summarises important information on the used model and its model parameters.

Table A.4: Overview of the applied NEQ model (Kloeker et al., 2005 and Buchaly et al., 2007)

Thermodynamics and physical properties			
VLE model	UNIQUAC (see Table A.2) and Hayden O'Connel		
LLE	UNIQUAC (see Table A.3)		
Reaction kinetics			
Catalyst	Amberlyst™ 46		
Kinetic model	Pseudohomogeneous approach		
Mass of dry catalyst per metre	0.205 kg/m		
Reactive section	2.1 m (Katapak™ SP 11)		
Axial discretisation			
Height of one discrete	HETP/4		
HETP of Sulzer BX™	0.14 m		

HETP of Katapak™–SP 11	0.5 m	
Hydrodynamics and mass transfer	Sulzer BX™	Katapak™–SP 11
Mass transfer coefficient	Bravo et al. (1985)	Brunazzi and Viva (2006)
Effective interfacial area	Bravo et al. (1985)	Brunazzi and Viva (2006)
Liquid holdup	Rocha et al. (1993)	Brunazzi and Viva (2006)
Pressure drop	Rocha et al. (1993)	Brunazzi and Viva (2006)

APPENDIX B. SUPPORTING INFORMATION

Operating Conditions of Experiment Exp1

Total feed rate [kg/h]	Molar feed ratio POH/ProAc. χ [-]	Reflux ratio RR [kg/kg]	Distillate-organic [kg/h]	Distillate-aqueous [kg/h]	Bottom [kg/h]
4.002	1.51	2.51	1.476	-	2.526

Liquid Phase Molar Composition and Vapour Phase Temperature Profile of Experiment Exp1

Height [m]	x_{POH} [mol/mol]	x_{water} [mol/mol]	x_{ProPro} [mol/mol]	x_{ProAc} [mol/mol]	T_v [K]
5.5	0.473	0.495	0.032	0.000	360.65
4.3	0.814	0.099	0.082	0.004	367.15

2.05	0.634	0.039	0.219	0.109	372.85
1.0	0.694	0.019	0.251	0.037	372.19
0.5	0.586	0.002	0.354	0.057	373.05
0	0.220	0.000	0.619	0.161	377.12
Reboiler	0.085	0.001	0.649	0.265	393.28

Operating Conditions of Experiment Exp2

Total feed rate [kg/h]	Molar feed ratio POH/ ProAc. χ [-]	Reflux ratio RR [kg/kg]	Distillate-organic [kg/h]	Distillate-aqueous [kg/h]	Bottom [kg/h]
4.001	1.51	2.51	1.258	-	2.743

Liquid Phase Molar Composition and Vapour Phase Temperature Profile of Experiment Exp2

Height [m]	x_{POH} [mol/ mol]	x_{water} [mol/ mol]	x_{ProPro} [mol/ mol]	x_{ProAc} [mol/ mol]	T_v [K]
5.5	0.419	0.550	0.031	0.000	359.84
4.3	0.805	0.116	0.079	0.000	366.55
2.05	0.598	0.048	0.225	0.129	372.80
1.0	0.677	0.014	0.255	0.054	371.96
0.5	0.650	0.003	0.294	0.053	372.67
0	0.384	0.000	0.502	0.114	374.69
Reboiler	0.192	0.000	0.572	0.236	387.15

Operating Conditions of Experiment Exp3

Total feed rate [kg/h]	Molar feed ratio POH/ ProAc. χ [-]	Reflux ratio RR [kg/kg]	Distillate-organic [kg/h]	Distillate-aqueous [kg/h]	Bottom [kg/h]
3.998	1.51	2.50	0.500	0.299	3.200

Liquid Phase Molar Composition and Vapour Phase Temperature Profile of Experiment Exp3

Height [m]	x_{POH} [mol/ mol]	x_{water} [mol/ mol]	x_{ProPro} [mol/mol]	x_{ProAc} [mol/ mol]	T_v [K]
5.5	0.292	0.587	0.121	0.000	358.70
4.3	0.468	0.409	0.123	0.000	359.75
2.05	0.519	0.069	0.270	0.142	372.77
1.0	0.627	0.032	0.287	0.054	371.25
0.5	0.644	0.000	0.298	0.059	372.35
0	0.545	0.000	0.376	0.079	373.55
Reboiler	0.379	0.000	0.454	0.166	379.32

Liquid Phase Molar Composition of Organic and Aqueous Distillate In Exp3

	x_{POH} [mol/ mol]	x_{water} [mol/ mol]	x_{ProPro} [mol/ mol]	x_{ProAc} [mol/ mol]
Distillate organic	0.353	0.495	0.152	0.000
Distillate aqueous	0.059	0.939	0.002	0.000

Operating Conditions of Experiment Exp4

Total feed rate [kg/h]	Molar feed ratio POH/ ProAc. χ [-]	Reflux ratio RR [kg/kg]	Distillate-organic [kg/h]	Distillate-aqueous [kg/h]	Bottom [kg/h]
4.000	1.51	2.69	0.730	0.200	3.071

Liquid Phase Molar Composition and Vapour Phase Temperature Profile of Experiment Exp4

Height [m]	x_{POH} [mol/ mol]	x_{water} [mol/ mol]	x_{ProPro} [mol/mol]	x_{ProAc} [mol/mol]	T_v [K]
5.5	0.301	0.593	0.106	0.000	358.73
4.3	0.419	0.497	0.072	0.012	360.16
2.05	0.630	0.049	0.201	0.121	371.37
1.0	0.665	0.017	0.264	0.055	371.33
0.5	0.663	0.000	0.281	0.057	372.44
0	0.522	0.000	0.395	0.083	373.61
Reboiler	0.360	0.000	0.460	0.180	380.70

Liquid Phase Molar Composition of Organic and Aqueous Distillate in Exp4

	x_{POH} [mol/ mol]	x_{water} [mol/ mol]	x_{ProPro} [mol/ mol]	x_{ProAc} [mol/ mol]
Distillate organic	0.328	0.554	0.118	0.000
Distillate aqueous	0.060	0.938	0.002	0.000

Operating Conditions of Experiment Exp5

Total feed rate [kg/h]	Molar feed ratio POH/ ProAc. χ [-]	Reflux ratio RR [kg/kg]	Distillate-organic [kg/h]	Distillate-aqueous [kg/h]	Bottom [kg/h]
4.000	1.51	3.99	0.286	0.340	3.374

Liquid Phase Molar Composition and Vapour Phase Temperature Profile of Experiment Exp5

Height [m]	x_{POH} [mol/mol]	x_{water} [mol/mol]	x_{ProPro} [mol/mol]	x_{ProAc} [mol/mol]	T_v [K]
5.5	0.291	0.590	0.119	0.000	358.84
4.3	0.504	0.401	0.095	0.000	360.40
2.05	0.557	0.079	0.235	0.129	371.88
1.0	0.645	0.020	0.278	0.057	371.25
0.5	0.658	0.000	0.286	0.056	372.55
0	0.570	0.000	0.359	0.071	373.65
Reboiler	0.397	0.000	0.444	0.159	379.32

Liquid Phase Molar Composition of Organic and Aqueous Distillate in Exp5

	x_{POH} [mol/mol]	x_{water} [mol/mol]	x_{ProPro} [mol/mol]	x_{ProAc} [mol/mol]
Distillate organic	0.350	0.502	0.149	0.000
Distillate aqueous	0.058	0.940	0.002	0.000

Operating Conditions of Experiment Exp6

Total feed rate [kg/h]	Molar feed ratio POH/ ProAc. χ [-]	Reflux ratio RR [kg/kg]	Distillate-organic [kg/h]	Distillate-aqueous [kg/h]	Bottom [kg/h]
3.998	1.51	2.36	0.905	0.145	2.949

Liquid Phase Molar Composition and Vapour Phase Temperature Profile of Experiment Exp6

Height [m]	x_{POH} [mol/mol]	x_{water} [mol/mol]	x_{ProPro} [mol/mol]	x_{ProAc} [mol/mol]	T_v [K]
5.5	0.303	0.603	0.093	0.000	358.65
4.3	0.432	0.526	0.042	0.000	360.02
2.05	0.477	0.092	0.309	0.122	372.36
1.0	0.663	0.021	0.264	0.052	371.44
0.5	0.661	0.003	0.282	0.054	372.16
0	0.511	0.000	0.402	0.087	373.36
Reboiler	0.327	0.000	0.495	0.178	381.14

Liquid Phase Molar Composition of Organic and Aqueous Distillate in Exp6

	x_{POH} [mol/mol]	x_{water} [mol/mol]	x_{ProPro} [mol/mol]	x_{ProAc} [mol/mol]
Distillate organic	0.324	0.575	0.101	0.000
Distillate aqueous	0.066	0.932	0.002	0.000

Operating Conditions of Experiment Exp7

Total feed rate [kg/h]	Molar feed ratio POH/ProAc. χ [-]	Reflux ratio RR [kg/kg]	Distillate-organic [kg/h]	Distillate-aqueous [kg/h]	Bottom [kg/h]
3.501	1.47	2.34	0.589	0.265	2.647

Liquid Phase Molar Composition and Vapour Phase Temperature Profile of Experiment Exp7

Height [m]	x_{POH} [mol/mol]	x_{water} [mol/mol]	x_{ProPro} [mol/mol]	x_{ProAc} [mol/mol]	T_v [K]
5.5	0.297	0.572	0.131	0.000	358.15
4.3	0.453	0.456	0.091	0.000	359.20
2.05	0.559	0.061	0.232	0.148	374.30
1.0	0.621	0.036	0.297	0.047	371.19
0.5	0.629	0.000	0.320	0.051	371.88
0	0.498	0.000	0.431	0.071	373.02
Reboiler	0.324	0.000	0.526	0.150	379.76

Liquid Phase Molar Composition of Organic and Aqueous Distillate in Exp7

	x_{POH} [mol/mol]	x_{water} [mol/mol]	x_{ProPro} [mol/mol]	x_{ProAc} [mol/mol]
Distillate organic	0.353	0.485	0.161	0.000

Distillate aqueous	0.059	0.939	0.002	0.000

REFERENCES

1.   Abrams, D.S., Prausnitz, J.M., 1975. Statistical thermodynamics of liquid mixtures: a new expression for the excess energy of partly or completely miscible systems. AIChE Journal 21, 116–128.

2.   Agreda, V.H., Partin, L.R., Heise, W.H., 1990. High-purity methyl acetate via reactive distillation. Chemical Engineering Progress 86, 40–46.

3.   Altman, E., et al., 2010. Pilot plant synthesis of n-propyl propionate via reactive distillation with decanter separator for reactant recovery. Experimental model validation and simulation studies. Chemical Engineering and Processing: Process Intensification 49, 965–972.

4.   Baur, R., et al., 2000. Comparison of equilibrium stage and nonequilibrium stage models for reactive distillation. Chemical Engineering Journal 76, 33–47.

5.   Bravo, J.L., Rocha, J.A., Fair, J.R., 1985. Mass transfer in gauze packings. Hydrocarbon Processing 64, 91–95.

6.   Brunazzi, E., Viva, A., 2006. Experimental investigation of reactive distillation packing Katapak-SP11: hydrodynamic aspects and size effects. In: Sørensen, E. (Ed.), Proceedings of the Conference Distillation and Absorption 2006. Institution of Chemical Engineers, Rugby, pp. 554–562.

7.   Buchaly, C., 2009. Experimental investigation, analysis and optimisation of hybrid separation processes. Dissertation, first. Verlag Dr. Hut.

8.   Buchaly, C., Kreis, P., Go´ rak, A., 2007. Hybrid separation processes-Combination of reactive distillation with membrane separation. Chemical Engineering and Processing: Process Intensification 46, 790–799.

9.   Chilton, T.H., Colburn, A.P., 1935. Distillation and absorption in packed columns. A convenient design and correlation method. Journal of Industrial and Engineering Chemistry 27, 255–260.

10. Duarte, C., et al., 2006. Esterification of propionic acid with n-propanol catalytic and non-catalytic kinetic study. Inzynieria Chemiczna I Procesowa 27, 273–286.

11. Forner, F., et al., 2008. Startup of a reactive distillation process with a decanter. Chemical Engineering and Processing: Process Intensification 47, 1976–1985.

12. Gmehling, J., et al., 2004. Azeotropic Data, 2nd ed. Wiley-VCH, Weinheim.

13. Go´ rak, A., Hoffmann, A., 2001. Catalytic distillation in structured packings: methyl acetate synthesis. AIChE Journal 47, 1067–1076.

14. Harmsen, G.J., 2007. Reactive distillation: the front-runner of industrial process intensification. A full review of commercial applications, research, scale-up, design and operation. Chemical Engineering and Processing: Process Intensification 46, 774–780.

15. Hayden, J.G., O'Connell, J.P., 1975. A generalized method for predicting second

16. virial coefficients. Industrial & Engineering Chemistry Process Design and Development 14, 209–216.

17. Hougen, O.A., 1947. Chemical Process Principles, 1st ed. Wiley, New York.

18. Kloeker, M., et al., 2005. Rate-based modelling and simulation of reactive separations in gas/vapour-liquid systems. Chemical Engineering and Processing: Process Intensification 44, 617–629.

19. Kloeker, M., et al., 2003. Influence of operating conditions and column configuration on the performance of reactive distillation columns with liquid–liquid separators. Canadian Journal of Chemical Engineering 81, 725–732.

20. Lai, I.-K., et al., 2008. Production of high-purity ethyl acetate using reactive distillation: experimental and start-up procedure. Chemical Engineering and Processing: Process Intensification 47, 1831–1843.

21. Lasdon, L.S., et al., 1978. Design and testing of a generalized reduced gradient code for nonlinear programming. ACM Transactions on Mathematical Software 4, 34–50.

22. Lewis, W.K., Whitman, W.G., 1924. Principles of gas absorption. Industrial and Engineering Chemistry Research 16, 1215–1220.

23. Luyben, W.L., 2008. Control of the heterogeneous azeotropic n-butanol/water distillation system. Energy and Fuels 22, 4249–4258.

24. Narasimhan, S., Jordache, C., 1999. Data Reconciliation and Gross Error Detection. An Intelligent use of Process Data, 1st ed. Gulf Professional Publishing, Houston.

25. NIST, 2009. Chemistry Web Book. /http://webbook.nist.gov/chemistryS (accesse December 2009).

26. Rocha, J.A., Bravo, J.L., Fair, J.R., 1993. Distillation columns containing structured packings: a comprehensive model for their performance. 1. Hydraulic models. Industrial and Engineering Chemistry Research 32, 641–651.

27. Schmitt, M., et al., 2004. Synthesis of n-hexyl acetate by reactive distillation. Chemical Engineering and Processing: Process Intensification 43, 397–409.

28. Schoenmakers, H., Beßling, B., 2003. Reactive and catalytic distillation from an industrial perspective. Chemical Engineering and Processing: Process Intensi- fication 42, 145–155.

29. Sørensen, J.M., Arlt, W., 1980. Ternary and Quarternary Systems. Tables, Diagrams and Model Parameters. DECHEMA, Frankfurt/M.

30. Steinigeweg, S., Gmehling, J., 2004. Transesterification processes by combination of reactive distillation and pervaporation. Chemical Engineering and Processing: Process

Intensification 43, 447–456.

31. Sundmacher, K., Hoffmann, U., 1996. Development of a new catalytic distillation process for fuel ethers via a detailed nonequilibrium model. Chemical Engineering Science 51, 2359–2368.

32. Sundmacher, K., Kienle, A. (Eds.), 2003. Status and Future Directions 1st ed. Wiley-VCH, Weinheim.

33. Taylor, R., 2006. (Di)still modelling after all these years: a view of the state of art. In: Sørensen, E. (Ed.), Proceedings of the Conference Distillation and Absorption 2006. Institution of Chemical Engineers, Rugby.

34. Taylor, R., Krishna, R., 2000. Modelling reactive distillation. Chemical Engineering cience 55, 5183–5229.

35. The Association of German Engineers, 2000. Uncertainties of measurement during acceptance tests on energy-conversion and power plants Fundamentals (VDI 2048 Part 1).

36. Widagdo, S., Seider, W.D., 1996. Azeotropic distillation. AIChE Journal 42, 96–130.

37. Wu, Y.-C., Chien, I.-L., 2009. Design and control of heterogeneous azeotropic column system for the separation of pyridine and water. Industrial and Engineering Chemistry Research 48, 10564–10576.

Chapter 5

Experimental Study of a Vane-Type Pipe Separator for Oil–Water Separation

Shi Shi-ying, Xu Jing-yu, Sun Huan-qiang, Zhang Jian, Li Dong-hui, and Wu Ying-xiang

Institute of Mechanics, Chinese Academy of Sciences, Beijing 100190, China

ABSTRACT

An experimental study of a new vane-type pipe separator (VTPS) was conducted for the possible application in the well-bore for oil–water separation and reinjection. Results by using particle image velocimetry (PIV) reveal a better flow field distribution for oil–water separation, which is formed in VTPS than that in hydrocyclone. The effects of split ratio, the oil content, guide vanes' installation and number of guide vanes on oil–water separation performance have been investigated experimentally. Compared to a traditional single hydrocyclone, VTPS shows a good separation performance as the water content at the inlet of VTPS reaches 79.9%, the oil content at the water-rich outlet is about 400 ppm while the split is near 0.70. These results are helpful to provide a possibly new design for downhole oil–water separation.

INTRODUCTION

When the produced water in mature oil fields continues to increase, it is of great significance to separate the ever-increasing volumes of water from oil downhole and inject it into a suitable formation. The adoption of these measures could not only extend the economic exploitation of oil fields, but also maintain the reservoir pressure (Chapuis et al., 1999).

The mentioned process is attractive but needs further investigation about the structure optimization of separator for the limited space in the well-bore. Traditional downhole separator is hydrocyclone which is a kind of tangential inlet structure combining with a long small cone tail and an upflow tube (Ogunsina and Wiggins, 2005). According to the number of tangential inlets, traditional hydrocyclone could be classified into two types: "multi-inlet" (Jiang et al., 2002) and "single-entry". While due to the well-bore's small diameter, hydrocyclone with a single-entry is chosen in most cases (Bowers et al., 1999). According to information on the downhole hydrocyclone installations in North America, it works in wells' diameter larger than 135 mm with water content of greater than 88.4% (Veil et al., 1999). These downhole applications present the limitation of hydrocyclone and there is not much experience with hydrocylone used on streams with high oil content as well. These restrictions are mainly due to the shortcomings of hydrocyclone's entry. The "single-entry" caused by the limited space is always small. Small inlets are more likely to cause oil droplets break-up (Listewnik, 1984) and thus exacerbate the difficulties of oil–water separation process (Meyer and Bohnet, 2003). Besides, the "single-entry" makes the structure of hydrocyclone asymmetric and so does the flow field which would cause the oil core start to weave, oscillate (Schutz et al., 2009) and re-mixing of oil droplets between the oil core and water to happen (Thew, 1986). If the oil phase is re-emulsified, it would be quite difficult to separate oil droplets from water and even lead to the presence of some oil droplets in water injected to a disposal zone. The potential problems of this lasting oily-water stream reinjection will add to the difficulties in subsurface injection of the produced liquid (van den Broek et al., 2001) and reduce the field oil production. To solve the above problems, Sooran et al. tried to improve the separation efficiency through the redesign of inlet structures. Michdet and Sangesland (1996) studied hydrocyclones with a small tube inside the underflow tube so as to recollect the oil phase in the underflow. However the quantity of water-removal would be lowered considering the fact that the underflow tube of hydrocyclone is already very small. Klasson et al. (2005) and Zhao et al. (2010) presented a new method to separate oil phase from water by getting rid of reverse flow in dynamic hydrocyclones. In the vortex flow field of dynamic hydrocyclones, water moves toward the wall, and then drains from the side outlet, while oil phase is forced to the center, and ejected through the centered oil outlet (Standridge et al., 1999). Their experiments showed that such improvement performed better than the traditional hydrocyclones. But as the downhole pressure is always very high, it is a problem to seal up the dynamic hydrocyclones and the separated-water is likely to be re-mixed with the liquid at the inlet.

The above literature review reveals a contradiction between the practical needs and limitation of hydrocyclones which is mainly caused by their inlet structure (tangential inlet in the narrow wells making their scale limited and that confining the efficiency) and their diverse flow direction between the inner and outer rotation flow. In view of this problem, inspired by the gas–liquid axial separator designed by Swanborn (Swanborn, 1988 and Delfos et al., 2004), a new kind of vane-type pipe separator (VTPS) is proposed in this work and the structure of a VTPS is shown in Fig. 1. The designs base on the idea of symmetrical flow field and decrease the crisis of oil phase to be re-injected. Thus the static guide vanes and tangentially drilled holes are designed. This VTPS for oil–water separation is a kind of axial flow cyclone and has some advantages in the aspects of separation performance as followings: Firstly, there is no slender cone and tangential inlet so that the VTPS has a low-pressure drop and a relative larger volume for oil–water separation (Standridge et al., 1999) which makes it quite suitable for the downhole oil–water separation; Secondly the diameter of tangentially inlet is always 0.14–0.17 times that of the cylinder so that the distance of an oil droplet moves from the inlet to the cylinder axis is short on the assumption of the same area of the cylinder's cross-section. This would shorten the time of the oil-droplet to move into the central area and increase the separation efficiency; Thirdly, in the VTPS, no reverse flow exists as well as no change in axial direction of flow between centrosymmetrical oil-core and the water phase near the wall. This reduces the risk of remix between the oil-core and surrounding water phase and help to promote the separation efficiency.

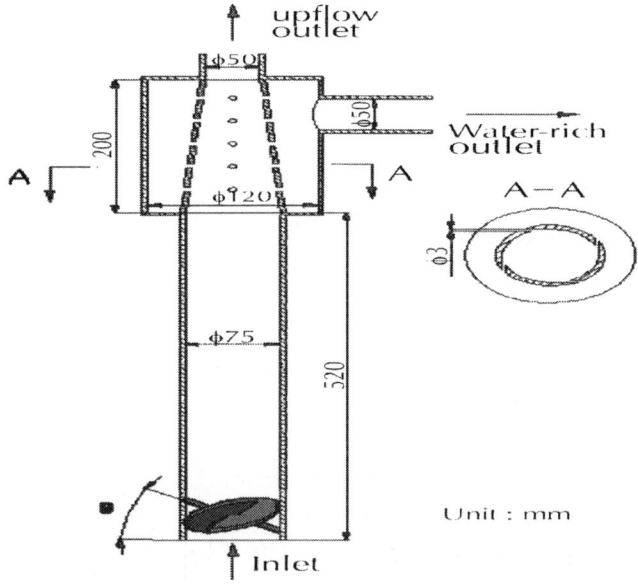

Figure 1: Schematic of VTPS.

EXPERIMENTAL

Experimental Setup and Operation Principle

An experimental flow loop has been constructed and shown in Fig. 2. The diameter of pipes which connect the VTPS and other installations in the flow loop are 50 mm. The geometrical details of the VTPS, made of plexiglass, are illustrated in Fig. 1, the pipe with the guide vanes installed near the inlet is 75 mm in diameter and 520 mm in length. At the other end of this pipe, a 200-mm long conical pipe, with four even distributed and tangentially holes in each cross section, connects it into the flow loop. The conical pipe is inside a 120 mm diameter concentric tube. A 50 mm diameter pipe located at the right top perpendicular to the concentric tube is used as the water-rich outlet.

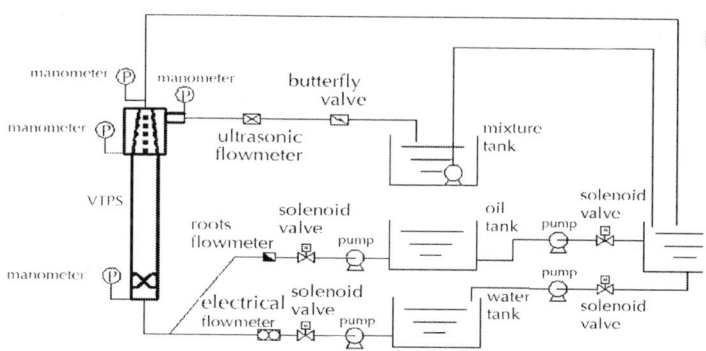

Figure 2: Flow loop of experimental system.

During the experiments, water and oil phases are first pumped from the water and oil tanks respectively, mixed at the Y-junction and then this mixture fluid is sent to the VTPS. Through the centrifugal separation in VTPS, the tangential velocity is gained by the guide vanes at the inlet and thus a central symmetry swirl motion is formed. In the swirl motion, oil and water phases are separated due to density difference. Oil phase is mainly distributed around the pipe central axis while water phase moves towards the wall. And then, a water-rich stream exits from the tangentially drilled holes into the water-rich outlet and an oil-rich stream ejects through the centered upflow outlet. Finally both phases are pumped to the mixture tank for gravity separation and re-circulation.

Control and Data Acquisition Systems

The LP-14 white oil and deionized water are used in the experiments, and their physical properties under test conditions (the atmosphere temperature is about 17 °C) are as follows: $\rho_o = 836.0$ kg/m^3, $\mu_o = 0.245$ kg/m s, $\rho_w = 1000.0$ kg/m^3, $\mu_w = 0.001$ kg/m s. The oil volume fraction at the inlet ranges from 2.0 to 20 percent.

The oil and water pumps are variable frequency pumps which could alter the running speed to control the flow rates of oil and water respectively. The flow rate at the water-rich outlet is changed by the butterfly valve and measured by an ultrasonic flowmeter. The flow loop is also equipped with several pressure transducers for pressure measurement. All output signals from the sensors are collected at a central data-acquisition panel. The oil concentration in the separated water phase is measured by OilTech 121A Handheld Oil in Water Analyzer with the measurement accuracy within 5 ppm.

Experimental Arrangement

For VTPS, the guide vanes are the key component to form swirl flow field. In this study, the effect of their installation angle and number n on oil–water separation performance is discussed. Table 1 shows the five designs and the thickness of guide vanes are 2 mm.

Table 1: Designs of different guide vanes for experiments

ϑ	20°	30°	40°
n	3	2, 3, 4	3

VELOCITY FIELD DISTRIBUTION

To measure the cross-sectional distribution of velocity field downstream of the guide vanes, a two-dimensional particle image velocimetry (2D PIV) is employed. The PIV system is produced by Germany LaVision company. The light source is a double pulsed Nd:YAG laser that sends short duration (4 ns) high energy (800 mJ) pulses of green light (532 nm). The collimated laser beam transmitted through one adjustable lens and two concave cylinder lens to generate a 1 mm thick lightsheet and illuminate the flow of a plane. The light reflected by the tracer particles (10 μm in average size, 1000 kg/m³) which is evenly distributed in water is recorded at 5 Hz by a CCD camera with 1376 × 1040 pixels. Images are formed through the imaging array. The analysis of the images is processed by adaptive correlation (Davis 7.2 Software) which gives a 64 × 64 vectors grid on 32 × 32 pixel-size final interrogation spots and the pixel resolution is 6.45 × 6.45 μm. Through the adaptive correlation, the location x(t), y(t) of the same tracer particle are functions of time t. So the velocity of water point that the tracer particle locates can be expressed as follows:

$$v_x = \frac{dx(t)}{dt} \approx \frac{x(t + \Delta t) - x(t)}{\Delta t} = \overline{v_x}$$

(1)

$$v_y = \frac{dy(t)}{dt} \approx \frac{y(t + \Delta t) - y(t)}{\Delta t} = \overline{v_y}$$

(2)

here v_x, v_y are the instantaneous velocity along the x direction and y direction respectively; $\overline{V}_x$, $\overline{V}_y$ are the mean velocity along the x direction and y direction respectively; Δt is the time interval between the two continuous shooting.

PIV system can obtain velocity vector of the entire planes and is proved to be an effective test method. Lim et al. (2010) adopted 2D PIV to study the velocity vector distributions within the hydrocyclone. Martins et al. (2010) compared the typical behaviour of the tangential and axial mean velocity components got by both LDA (Laser Doppler Velocimetry) and 2D PIV. They found a good agreement between the two methods.

During this experiment, to minimize the effects of reflection and refraction of the light beams, a water cube tank is placed surround the VTPS. The averaged (100 samples in 20 s) flow field of the cross section and axis cross-section in a VTPS with 4 guide vanes at   of 30° is shown in Fig. 3. The axial velocity profile could be obtained is from the averaged flow field of axis cross-section, while the tangential velocity profiles is acquired from averaged flow field of cross-section.

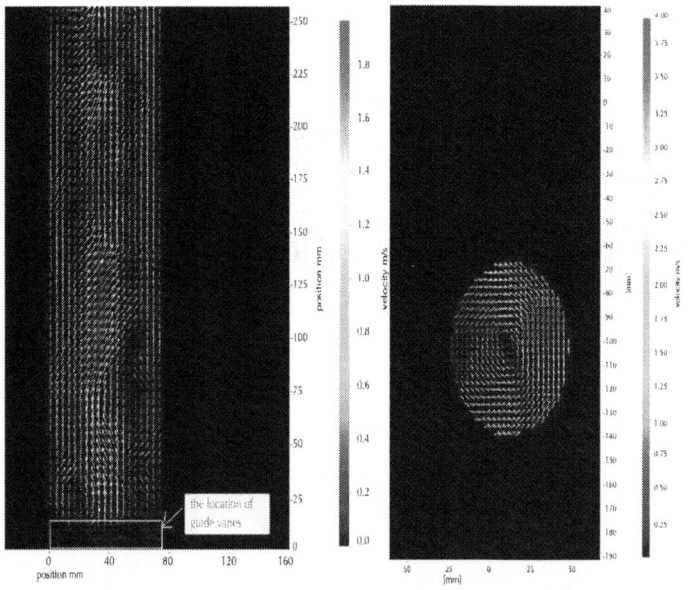

Figure 3: The averaged velocity field of the axis cross section (left) and cross section (right) in the VTPS.

Fig. 4 gives the axial and tangential velocity profile in the VTPS where the axial distance from the guide vanes is double of the pipe's diameter when the flowrate of water at the inlet is 12.00 m³/h. The axial velocity has no zero vertical velocity, which means that there does not exist axial reverse flow. Thus, this structure reduces the remix between the oil core in the center area and water phase in the surrounding area. Another characteristic of axial velocity profile is that the maximum value appears in the center area, which helps to prevent the oil droplet in the center area from flowing into the water-rich stream in the cone section. However, in the hydrocyclone, the axial velocity near the wall is negative and positive in the center area. This induces that the maximum axial velocity in the center area is larger in the VTPS than the hydrocyclone. Therefore, the VTPS performs better than the hydrocyclone for its rapid moving the oil phase into the center area. From the tangential velocity distribution, it is a central symmetric vector field. In a word, the flow field distribution in the VTPS is better for oil–water separation.

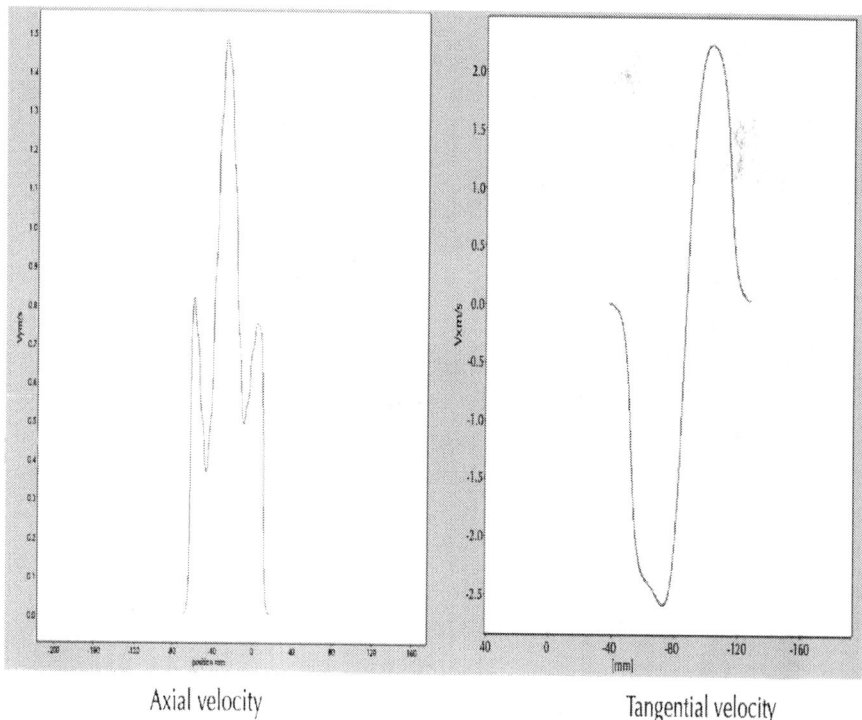

Axial velocity   Tangential velocity

Figure 4: The tangential and axial velocities profiles in the VTPS.

RESULTS OF SEPARATION PERFORMANCE AND DISCUSSION

Effects of Split Ratio

To investigate the effect of the split ratio on oil–water separation in the VTPS, experiments were carried out with θ of 20° and the same guide vanes' number of 3.

Split ratio is defined as:

$$F = \frac{Q_w}{Q_i} \tag{3}$$

where Q_w is the flowrate at the water-rich outlet, Q_i represents the flowrate at the inlet.

Fig. 5 shows the effect of split ratio on the distribution of oil and water phases in the VTPS. The gray in the center of VTPS represents oil phase. When the flowrate is 3.53 m³/h, the oil content is 8.0% at the inlet. It can be seen that as the increasing of split ratio, the diameter of oil core in the VTPS enlarges. Fig. 6 gives the effect of split ratio on both the oil content at the water-rich outlet and the water content at the upflow outlet. The curve in diagram shows that the oil content at the water-rich outlet increases with the increasing of split ratio. It means that there exists a critical split ratio to reinject this water-rich mixture within the reinjection standard into the ground. In this experiment, the water content at the upflow outlet reaches 40.1% with oil concentrations in the separated water phase of less than 250 ppm. This value shows that the VTPS performs better than the typical hydrocyclone in which the water content at the overflow of 50.0–67.0% with oil concentration in the separated water phase of 100–500 ppm (Klasson et al., 2005).

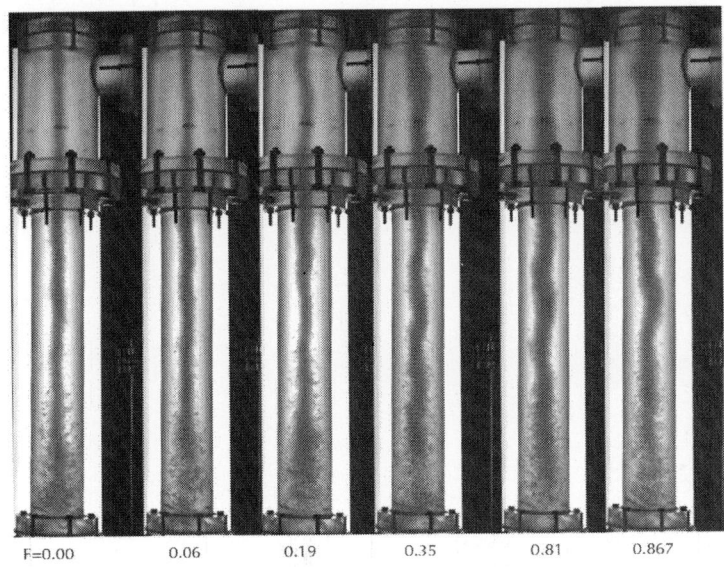

Figure 5: Distribution of oil and water in the VTPS under different split ratio.

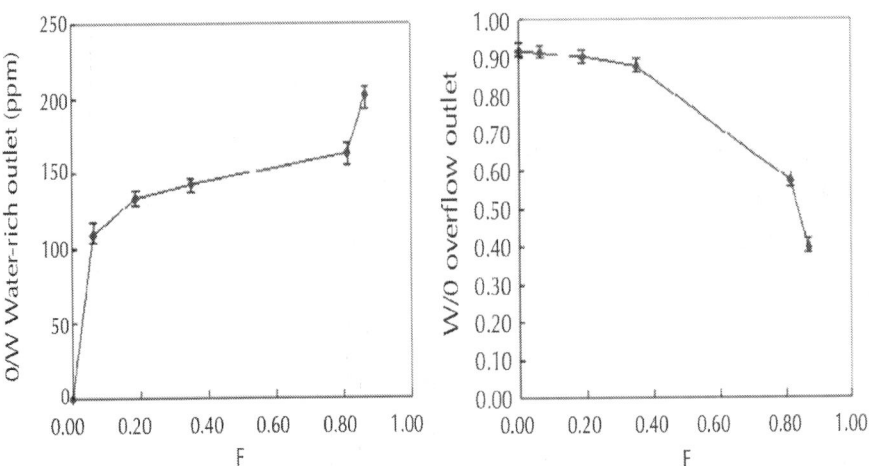

Figure 6: Effect of split ratio on oil–water separation performance.

Effects of the Oil Content

Fig. 7 shows the effect of oil content on purifying in VTPS with of 20° and the same guide vanes' number of 3 under different split ratio. Fig. 8 shows the effect of oil content on distribution of oil and water in the VTPS at zero split ratio.

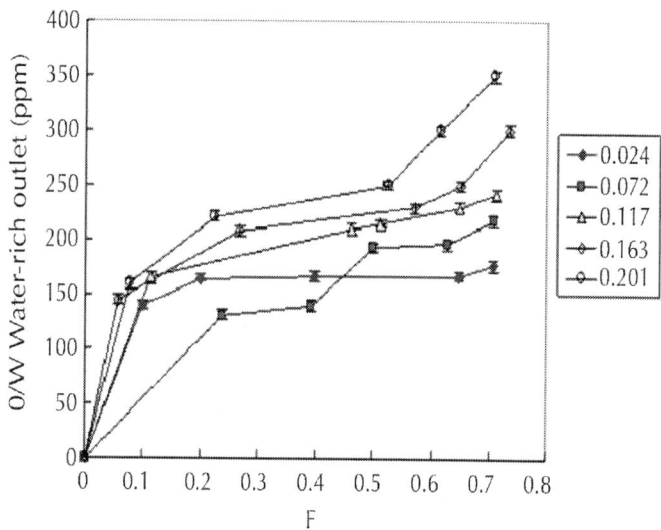

Figure 7: Effect of oil content on purifying in the VTPS under different split ratios.

As shown in Fig. 7, when the flowrate at the inlet is 4.07 m³/h and the split ratio is smaller than 0.40, with the increase of the oil content at the inlet, the oil content at the water-rich outlet decreases firstly, and then increases. According to the experimental observation shown in Fig. 8, when the oil content is small, the oil phase entering the inlet of the VTPS are dispersed more small balls. When the oil content increases, the oil phase is scattered as churns and the average diameter of droplet after the break-up in VTPS is larger. So the oil content at the water-rich outlet decreases first and then increases because the diameter of oil core increases quickly as the oil content at the inlet is greater than 6.0%.

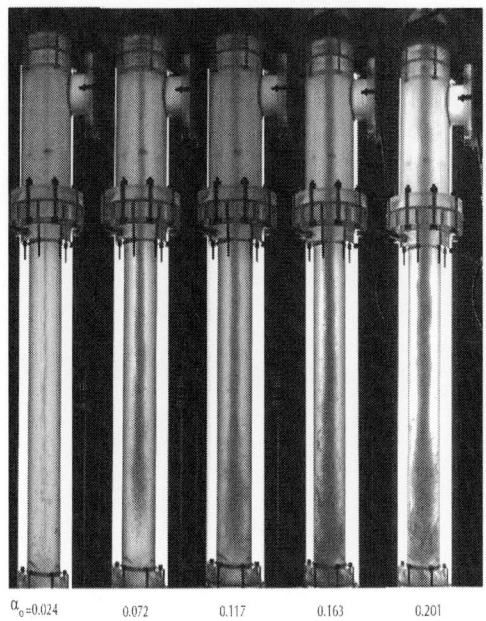

$\alpha_o = 0.024$ 0.072   0.117   0.163   0.201

Figure 8: Distribution of oil and water at different oil contents.

From Fig. 7 and Fig. 8, it is observed that when the oil content arrived at 20.1%, the oil content at the water-rich outlet is within 400 ppm with the split ratio of 0.70. It means that the VTPS could be used in wells with water content of 79.9% which seems an exciting result compared to the hydrocyclone using in wells with water content of more than 88.4%.

Effects of Guide Vanes' Installation Angle

For the VTPS, the installation angle of guide vanes is another important geometric parameter for influences on the oil–water separation. To illustrate this, some experiments were discussed under the same flow rate at the inlet and the same guide vanes' number of three in the VTPS.

Fig. 9 shows the effect of guide vanes' installation angle on the oil and water distribution in the VTPS with zero split ratio.

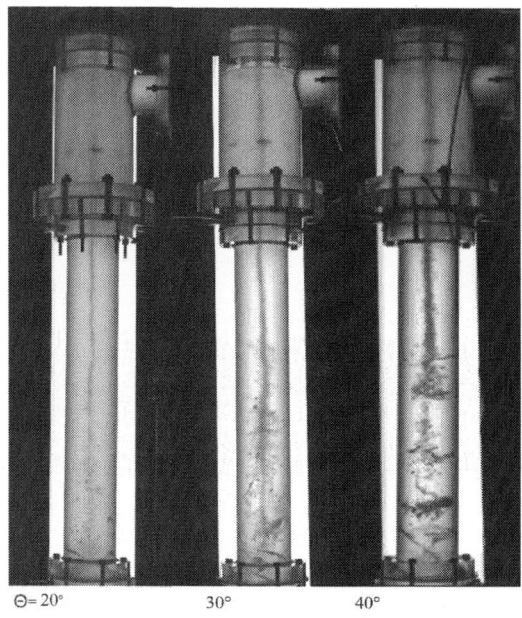

Figure 9: Distribution of oil and water in the VTPS with different guide vane's installation angles.

It can be found that when the flowrate is 4.07 m³/h and the oil content is 2.4% at the inlet, as the installation angle increases, the average droplet diameter in VTPS increases and so does the diameter of oil core.

Fig. 10 shows the effect of different guide vanes' installation angle on purifying after oil–water separation in VTPS with different split ratio. It could be observed that when the split ratio is below 0.10, the installation angle of 40° has the best result because the concentration of oil in water at the water-rich outlet is the smallest among these three designs. When the split ratio is large than 0.10, the VTPS with installation angle 30° has the least oil cut at the water-rich outlet. From Fig. 9, it could be found that at the same condition, as the decrease of the guide vane's angle, the smaller the oil droplet increased. The droplet in VTPS is too small to be separated from water. When the guide vane's angle is too large, the tangential velocity diverted is not very large. Although the oil droplet is large, the vortex formed by the tangential velocity is not strong enough to separate the mixture fluid. So when the flow rate at the inlet is 4.00 m³/h, the VTPS with guide installation angle of 30° is the best.

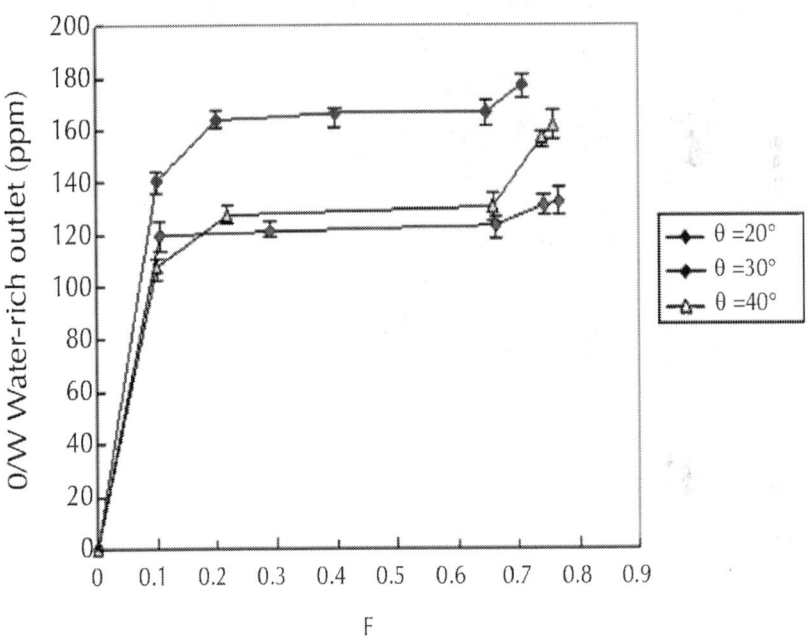

Figure 10: Effect of guide vanes' installation angle on purifying.

Effects of Guide Vanes' Number

Fig. 11 and Fig. 12 show the number of guide vanes' effect on oil–water separation in VTPS with the same of 30°. When the mixture flowrate at the inlet is 7.30 m³/h and the oil volume fraction is kept at 2.4%, the distribution of oil and water phases in VTPS with F = 0 is presented in Fig. 11. It indicates that as the number of guide vanes increases, the axial distance for the oil droplets moving to the centre area is shorter.

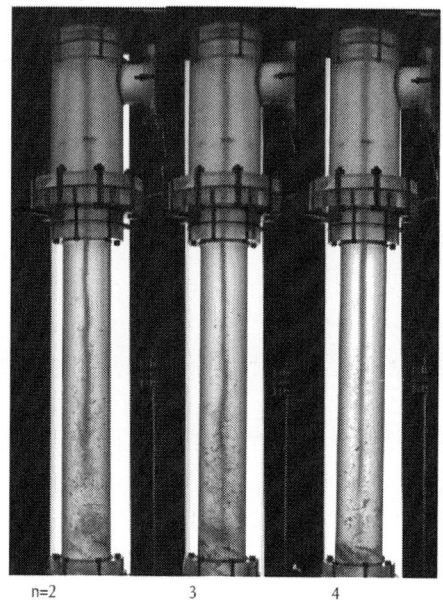

Figure 11: Distribution of oil and water in the VTPS with different guide vane's numbers.

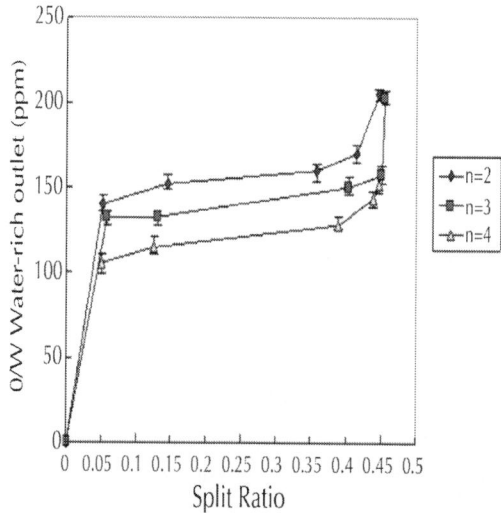

Figure 12: Effect of guide vanes' number on purifying.

Fig. 12 shows the oil volume fraction at the water-rich outlet of VTPS with different guide vanes' number when the split ratio increases. As can be observed that the oil content at the water-rich outlet decreases with the increases of guide vanes' number under the same split ratio. Increasing the number of guide vanes means more mixture fluid converted by the guide vanes to improve the average tangential velocity of the fluid. The force on a small droplet in this flow could be assumed to abide by Stokes law (Bird et al., 1960). The radial velocity is calculated through a balance between the Stokes drag and the centrifugal force on another reasonable assumption that the dispersed phase has the same axial and tangential velocity as the continuous phase (Frans et al., 1995):

$$\frac{4(\rho_w - \rho_0)\pi D^3 v_\theta^2}{3\pi} = 6\pi v \rho_0 D \frac{dr}{dt} = 6\pi v \rho_0 D v_r$$

(4)

where D is the nominal diameter of oil droplet; ρ_w and ρ_0 are the densities of oil droplet and water; v_θ is the tangential velocity; r is the revolving radius of the droplet in VTPS; v is the kinematic viscosity; v_r is the radial velocity. It is implied that under the same other conditions, the radial velocity of the oil droplet is proportional to the tangential velocity and the size of the droplet. When the inlet flow rate is not very large, the more the guide vanes, the more strong tangential flow are converted from the axial flow at the inlet. As a result, the average tangential velocity is larger and the radial velocity of the oil droplet moving is faster. The oil droplet takes shorter time to move into the central area and the diameter of oil droplet which migrates into the central area within the same time is smaller. At the same rest conditions, as the number of guide vanes increases, the oil volume fraction at the water-rich outlet decreases.

CONCLUSIONS

In this work, two-dimensional flow in a new type of oil–water separator (VTPS) for the possible downhole applications has been characterized through the PIV techniques. The characteristics of tangential and axial velocity profiles have been acquired and show a new thought of oil–water separator.

The water content at the oil-rich stream after the separation in VTPS can reach is lower than the typical hydrocyclone, thus the separation efficiency is higher. Besides, VTPS has a wide working oil content range up to 20.1% and compact structure which has the possibility to handle a large amount of produced liquid downhole.

As the increase of split ratio, the oil-content in water-rich stream raises and is beneficial to the oil–water separation in VTPS. But too large a split ratio will lead to a certain amount of oil to be drained in to ground, so a critical split ratio exists when the oil-content in the water-rich stream reaches the reinjection standard.

The installation angle and number of guide vanes affects the diameter of the oil droplets distribution and the converted tangential velocity in VTPS. Under the same testing conditions and the designs in this work, installation angle of 30° is the optimum and four guide vanes are the best.

REFERENCES

1. Bird, R.B., Stewart, W.E., Lightfoot, E.N., 1960. Transport Phenomena. Wiley, New York, 59.

2. Bowers, B.E., Brownlee, R.F., Schrenkel, P.J., 1999. Development of a downhole oil/water separation and reinjection system for offshore application. SPE Prod. Facilities 15 (2), 115–122.

3. Chapuis, C., Lacourie, Y., Lancois, D., 1999. Testing of Down Hole Oil/Water Separation System in Lacq Superieur Field. France, SPE Paper, No. 54748.

4. Delfos, R., Murphy, S., Stanbridge, D., Olujic, Z., Jansens, P.J., 2004. A design tool for optimising axial liquid–liquid hydrocyclones. Miner. Eng. 17, 721–731.

5. Frans, T.M., Nieuwstadt, Dirkzwager, Maarten, 1995. Fluid

mechanics model for an axial cyclone separator. Ind. Eng. Chem. Res. 34, 3399–3404.

6.   Jiang, M.-h., Zhao, L.-x., Wang, Z., 2002. Effects of geometric and operating parameter on pressure drop and oil–water separation performance for hydrocyclones. In: Proceeding of

7.   The Twelfth International Offshore and Polar Engineering Conference, Kitakyushu, Japan, May 26–31, pp. 102–106.

8.   Klasson, K. Thomas, Taylor, Paul A., Walker Jr., Joseph F., Jones, Sandie A., Cummins, Robert L., Richardson, Steve A., 2005.

9.   Modification of a centrifugal separator for in-well oil–water separation. Sep. Sci. Technol. 40, 453–462.

10.   Lim, Eldin Wee Chuan, Chen, Y.-r., Wang, C.-h., Wu, R.-m., 2010. Experimental and computation studies of multiphase hydrodynamics in a hydrocyclone separator system. Chem. Eng. Sci. 65, 6415–6424.

11.   Listewnik, J., 1984. Some Factors Influencing the Performance of De-Oiling Hydrocyclone for Marine Applications.Second International Conference on Hydrocyclones, England, September 19–21.

12.   Martins, L.P.M., Duarte, D.G., Loureiro, J.B.R., Moraes, C.A.C., Silva Freire, A.P., 2010. LDA and PIV characterization of the flow in a hydrocyclone without an air-core. J. Petrol. Sci. Eng. 70, 168–176.

13.   Matthias Meyer, Matthias Bohnet, 2003. Influence of entrance droplet size distribution and feed concentration on separation of immiscible liquids using hydrocyclones. Chem. Technol. 26, 660–665.

14.   Michdet, J.F., Sangesland, S., 1996. Downhole separation of oil and water. In: The 9th Underwater Technology Conference, Bergen, Norway, March 20–22.

15.   Ogunsina, O.O., Wiggins, M.L., 2005. A Review of Downhole Separation Technology. SPE Paper, No. 94276.

16.   Schutz, S., Gorbach, G., Piesche, M., 2009. Modeling

fluid behavior and droplet interactions during liquid–liquid separation in hydrocyclones. Chem. Eng. Sci. 64, 3935–3952.

17.   Sooran, N., Hassan, S., Hashemabadi, 2009. CFD simulation of inlet design effect on deoiling hydrocyclone separation efficiency. Chem. Eng. Technol. 32 (12), 1885–1893.

18.   Standridge, D., Swanborn, R., Olujic, Z., 1999. A novel recycle axial flow cyclone with strongly improved characteristics for high-pressure and high-throughput operation. Multiphase, 555–563.

19.   Swanborn, R.A., 1988. A new approach to the design of gas–liquid separators for the oil industry, Ph.D. Thesis, Delft University of Technology.

20.   Thew, M.T., 1986. Hydrocyclone redesign for liquid–liquid separation. Chem. Eng., 17–23.

21.   van den Broek, W.M.G.T., van der Zande, M.J., Janssen, P.H., 2001. Downhole dehydration vs.reduction of oil droplet break-up. SPE Paper, No. 66542.

22.   Veil, John A., Langhus, Bruce G., Belieu, Stan, 1999. DOWS reduce produced water disposal costs. Oil Gas J. March, 76–85.

23.   Zhao, L.x., Li, F., Ma, Z.-z., Hu, Y.-q., 2010. Theoretical analysis and experimental study of dynamic hydrocyclones. ASME 132 (042901), 1–6.

Hydrodynamics and Velocity Measurements in Gas–Liquid Swirling Flows in Cylindrical Cyclones

Rainier Hreiz, Caroline Gentric, Noël Midoux,
Richard Lainé and Denis Fünfschilling

LRGP, CNRS-UMR 7274, Université de Lorraine, ENSIC, 1 rue Grandville, 54001 Nancy, France

ABSTRACT

The gas–liquid swirl flow in a gas–liquid cylindrical cyclone separator has been characterized first qualitatively by flow visualizations. The emerged findings were then confirmed quantitatively by Laser Doppler Velocimetry measurements. The vortex core presents a very complex hydrodynamics, characterized by an alternation between a laminar and a turbulent state. The laminar regime is associated with velocities pointing in the same direction as the mean flow, while the turbulent state induces velocities in the opposite direction, i.e. a flow reversal. These observations give a first understanding of the origin of the double flow reversal regime that is encountered in swirl flows. It is shown that this flow structure appears for high swirl intensities, and results from a frequent laminarization of the vortex core. Results show that, contrary to the commonly assumed hypothesis, this flow structure is associated with good separation performance of the cyclone. Accordingly, we propose the use of multiple tangential inlets to generate the swirl motion in the cyclone, which is supposed to favor the double flow reversal regime, and thus, improve the separation efficiency.

INTRODUCTION

Swirl flows are generated by giving a tangential velocity component to an axial flow, which results in a helical winding of the streamlines. The research in the domain has been motivated by two somewhat antagonist reasons: on one hand, phase separation by centrifugation, and on the other hand, the improvement of mixing and transport phenomena. Swirl flows are nowadays used in a broad range of engineering applications, with various goals: phase separation in cyclone separators (Rosa et al., 2001andTue-Nenu and Yoshida, 2009), mass transfer improvement (Yapici et al., 1997), heat transfer enhancement (Martemaniov and Okulov, 2004), reduction of localized wear in hydraulic or pneumatic transport (Fokeer et al., 2009), flame stabilization in burners (Wegner et al., 2004), etc.

This paper will focus on swirl flows in the context of the GLCC©,[1]a gas–liquid cyclone separator. The results concern swirl flows in a more general and broader context, however, it should be kept in mind that some characteristics of swirl flows depend on the method used to generate the swirl motion (Kitoh, 1991andMartemaniov and Okulov, 2004).

The GLCC (Fig. 1) is a gas–liquid separator of great interest for the oil industry, and that follows the "reverse-flow cyclone" technology. The GLCC is simple, compact, low-cost and low-weight compared to conventional separators, and contains no moving parts nor internal devices (so it requires only little maintenance). For these reasons, the use of the GLCC allows considerable saving in offshore, arctic, subsea and downhole operations.

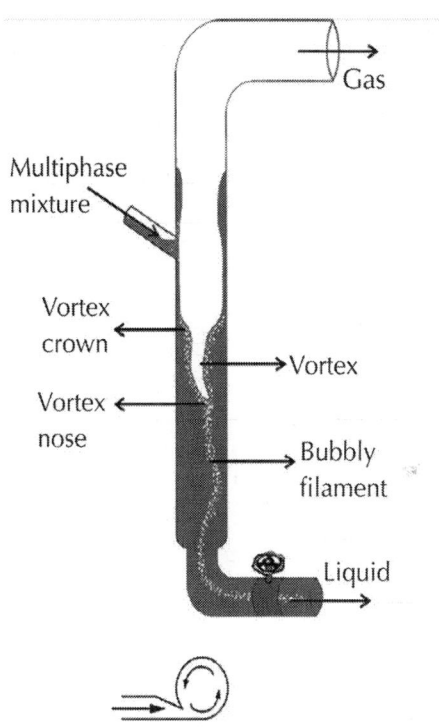

Figure. 1: Schematic representation of the GLCC and of the nozzle at its entrance pipe.

The GLCC consists of a vertical pipe with a downward inclined tangential inlet located approximately at mid-height of the cyclone body, and two outlets, respectively, at the top and bottom of the pipe. The inlet generally ends with a nozzle: thus the mixture is injected into the cyclone body at a higher velocity, which increases the centrifugal effects in the flow and enhances the phase separation process. During regular operation, the gas exits from the top of the GLCC while the liquid is collected from the bottom outlet.

Once the mixture enters the GLCC, because of gravity, a first phase separation occurs: the liquid tends to move toward the cyclone lower part, while gas occupies the upper part. Liquid height in the GLCC is controlled through automated valves mounted at the exits (seeFig. 1) (Wang, 2000). Liquid height must not exceed the inlet level to avoid liquid getting sprayed by the gas stream and being carried over to the gas outlet, and must be high enough to ensure sufficient residence time to separate bubbles from liquid. Thus, liquid level control is crucial for optimal performance of the GLCC under variable operating conditions, and should enable the GLCC to better handle surging and slugging events (Shoham and Kouba, 1998).

In the GLCC upper part, liquid droplets contained in the gas stream are centrifuged toward the walls, and coalesce into a liquid film. As this film is compact compared to individual droplets, the gas will have more difficulties to drive it up to the top outlet. The liquid from the wall film falls down by gravity into the liquid vortex thereafter, unless the liquid and gas flow rates couple exceeds the limit tolerated by the system. In this last situation, gas drags liquid from the film toward the GLCC upper outlet: this limiting phenomenon is called Liquid Carry-Over (LCO) (Hreiz et al., 2014). To improve the separator performance, it is recommended to incline the inlet downward by an angle of about 27° from the horizontal (Kouba et al., 1995). Compared to a horizontal inlet, this inclination directs the liquid stream below the inlet, preventing it from blocking the gas passage. It also promotes a stratified regime at the inlet (and thus phase segregation) and disadvantages the slug regime, prejudicial to the separation.

In the GLCC lower part, if the swirl intensity is high enough, the free gas–liquid interface gets carved out and the liquid vortex can be observed. The liquid vortex will be referred simply to as the vortex. The liquid flows from the inlet nozzle to the vortex in a thin swirling film, dragging down gas bubbles. In the vortex, large bubbles quickly move toward the free interface due to buoyancy and disengage. Smaller bubbles are dragged downward by the liquid, while getting pushed radially toward the vortex center due to centrifugal forces. They form a bubbly filament which allows a nice visualization of the vortex core. These bubbles are then supposed to rise up to the free interface and to disengage (see Section2.2.2for details). However, a fraction of the small gas bubbles passes along with the liquid underflow from the GLCC. This undesired phenomenon is called Gas Carry-Under (GCU) (seeHreiz et al. (2014)for details). The GLCC is nowadays mainly used to control the gas/liquid ratio upstream of equipments such as pumps, flow meters or desander hydrocyclones: this enhances their performance, and reduces their size and cost (Shoham and Kouba, 1998). Other applications of the GLCC as portable well testing equipment, pre-separator or partial separator have been reported (Shoham and Kouba, 1998). However, despite its significant potential, the GLCC as a full gas–liquid separator has not reached yet a widespread deployment: industry is still suspicious about its design. In fact, albeit the simplicity of the unit, the hydrodynamics in the cyclone is very complex, and many of the phenomena taking place are still not fully understood. As a consequence, the GLCC design is still empirical, so scale up of the laboratory-scale model to real industrial prototypes does not rely on solid and reliable basis.

The aim of this paper is to better understand the swirl flow hydrodynamics in the GLCC lower part by means of experiments on a laboratory-scale model. Dispersed gas bubbles act as a tracer allowing visualization of the vortex core. Key phenomena governing the behavior of the vortex core are identified through flow visualization. These observations reveal a different flow behavior from that described in the literature. The emerging conclusions are supported quantitatively by Laser Doppler Velocimetry (LDV) measurements in the GLCC lower part. This data permit to fill partially the lack of reported velocity measurements in gas–liquid swirl flows. Finally, based on these outcomes, a modified GLCC geometry that is expected to enhance the cyclone efficiency is proposed.

HYDRODYNAMICS IN THE GLCC LOWER PART

Global Hydrodynamic Behavior: Vortex Flow Patterns

Hreiz et al. (2014)studied the gas–liquid vortex flow patterns in the GLCC lower part (the same experimental setup is used in the current study (Section3.1)). Their results revealed that the vortex pattern depends mainly on the liquid flow rate and on the liquid height in the GLCC (controlled by a valve mounted at the lower outlet). An increase in the gas flow rate has little effect on the vortex pattern, but increases the bubble density. Apart for very low liquid flow rates, part of the bubbles rejoins the vortex center due to the centrifugal force and forms a bubbly filament, thus allowing to visualize the vortex core. The hydrodynamic behavior of the bubbly filament is discussed in Section4.1.

The*gravity dominated*and the*bubbly vortex*(Fig. 2a) flow patterns occur for low liquid flow rates. The associated centrifugal forces are weak, so the free interface is almost flat. They are characterized by moderate to high bubbles dispersion. For higher liquid flow rates, due to higher swirl intensities, the free interface hollows and the*excavated vortex*(Fig. 2b) flow pattern occurs. Under this flow regime, bubbles dispersion and GCU decrease with the increase of the liquid flow rate. For higher liquid flow rates, the vortex becomes*deeply excavated*(Fig. 2c) and shows tortuosities. The warping of the vortex comes from the use of a unique inlet what induces a pronounced asymmetry in the flow. This vortex pattern is associated to intense swirl intensities. Extremely few bubbles are found outside the region around the bubbly filament and the zone near the free interface. Their size is very small, and they are not clearly visible to the naked eye. Low GCU levels are observed under this flow regime.

Figure. 2: Different vortex regimes in the lower part of the GLCC (cases where the vortex crown is 21cm below the inlet level) (Hreiz et al., 2014). (a) Bubbly vortex flow ($V_{s,c,l}$=0.194m/s, $V_{s,c,g}$=0m/s). (b) Excavated vortex flow ($V_{s,c,l}$=0.309m/s, $V_{s,c,g}$=2.274m/s). (c) Deeply excavated vortex flow ($V_{s,c,l}$=0.584m/s, $V_{s,c,g}$=0.853m/s).

Note that lowering the liquid height in the cyclone promotes low swirl intensities flow patterns which occur then for a broader range of liquid flow rates. Details on the global hydrodynamic behavior in the GLCC lower part can be found inHreiz et al. (2014) orHreiz (2011).

Local Hydrodynamic Behavior

To the authors' knowledge, no local velocity measurements in such gas–liquid swirl flows have been reported so far in the literature. Therefore, this review is based on data reported in works done with single-phase swirl flows in straight pipes (we mean by single-phase swirl flows, internal swirl flows in pipes, i.e. without a free interface), in particular (Algifri et al., 1988,Kitoh, 1991,Chang and Dhir, 1994andErdal and Shirazi, 2002). OnlyErdal and Shirazi (2002)generated their swirl flows using a unique tangential inlet (as in the GLCC), what induced a pronounced vortex warping and resulted in 3D flow field.Chang and Dhir (1994)used 4 or 6 horizontal tangential injectors located at a same height.Algifri et al. (1988)andKitoh (1991)used radial cascade blades and guide vanes, respectively, as swirl generators. In these three works, the resulting swirl flow can be approximated as axisymmetric given the smooth conditions of swirl generation.

Mean Tangential Velocity Profile

A typical mean tangential velocity profile in swirl flows (Algifri et al., 1988,Kitoh, 1991,Chang and Dhir, 1994andErdal and Shirazi, 2002) is plotted inFig. 3. It can be divided in several regions:

A central region: The mean tangential velocity profile corresponds to a forced vortex, i.e. a solid body rotation. According to the Rayleigh criterion for stability to small perturbations ($r^{-3}[d(rW)^2/dr]>0$, where$W$is the mean tangential velocity and$r$is the radial position in cylindrical coordinates), this velocity distribution has a strong stabilizing effect, so the small scale turbulent motion dies rapidly.

An annular region: The mean tangential velocity approximates a free-vortex-type distribution. The turbulent motion contains the flow skewness effect in addition to the curvature effect (Kitoh, 1991). The free vortex zone is separated from the core region by a transition zone that can be quite large.

A wall region: It consists in a small layer near the wall, where the tangential velocity decreases with a steep gradient to be equal to zero at the wall. Turbulence is affected by the streamline curvature, which results in a deviation from the universal logarithmic profile (Kitoh, 1991).

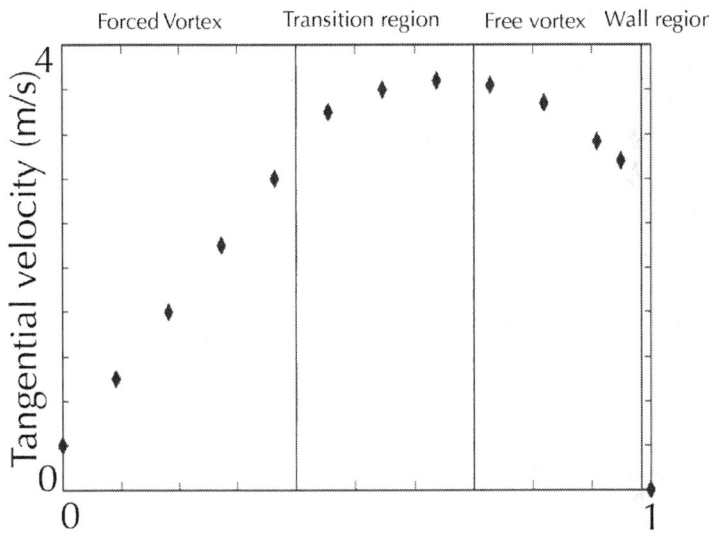

Figure. 3: Typical average tangential velocity profile in swirl flow, plotted on pipe radius R (r being the radial coordinate).

Mean Axial Velocity Profile

Based on experimental data from the literature obtained on single-phase swirl flows in pipes, Mantilla understood that the axial velocity profile in the GLCC lower part can be described by one of the three flow regimes presented inFig. 4(Mantilla, 1998).

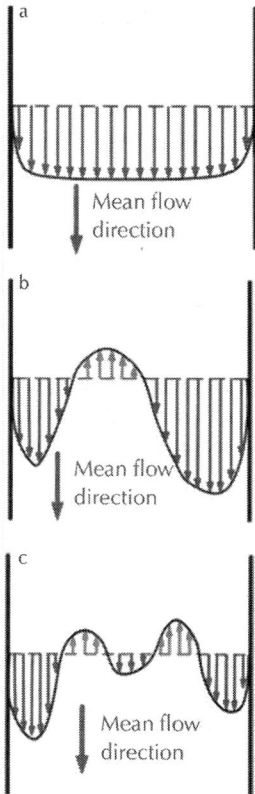

Figure. 4: Possible average mean axial velocity profiles in the GLCC. Note that the flow asymmetry will be less pronounced if multiple inlets are used. (a) Profile 1, forward flow over the entire flow section. (b) Profile 2, including a single flow reversal. (c) Profile 3, including a double flow reversal.

The first profile (Fig. 4a) corresponds to a very weak swirl intensity. It presents a quasi-uniform forward axial velocity on the entire cross-section.

The second velocity profile (Fig. 4b) corresponds to an important swirl intensity. If the swirl intensity is high enough, the flow reverses near the vortex center. In fact, due to the centrifugal forces, a radial pressure gradient develops and results in a low pressure region around the vortex center. Going downstream, the pressure in the core region increases due to swirl decay. The resulting adverse pressure gradient in the axial direction, if severe enough, reverses the flow direction.

The third velocity profile (Fig. 4c) corresponds also to an important swirl intensity. It presents a double flow reversal, with forward flow near the vortex center and the wall, and backward flow in the intermediate region (we will use the term "double flow reversal" even though it is not very accurate; in fact, the flow reversal region has an annular shape). References about this third flow regime are very few, and it has not been encountered in swirl flows generated by a unique tangential inlet. It has been reported in the works ofGuo and Dhir (1990)andNissan and Bresan (1961), where multiple tangential inlets were employed to generate the swirling motion. Later on, LDV measurements byErdal and Shirazi (2002)revealed the existence of a flow field of type 3 in a high Reynolds (Re=66 855) single-phase swirl flow generated by two diametrically opposed tangential inlets. For the same flow rate, but with a single inlet of similar total injection area (because of manufacturing difficulties, the single inlet has a total area 1.4 larger than the two inlet configuration), the mean axial velocity profile was of type 2.

For good separation efficiency, GLCC operation requires relatively high swirl intensities. That means that the encountered velocity fields in GLCCs are either type 2 or 3. As the GLCC is fed with a unique tangential inlet, and given that the axial flow field of type 3 has only been encountered in swirl flows generated by multiple tangential inlets, Mantilla deduced that the velocity field of type 2 is more likely to be found in GLCC separators (Mantilla, 1998).

Flow field of type 2 is considered beneficial for bubbles separation efficiency in the GLCC. In the GLCC lower part, centrifugal forces push the bubbles radially toward the vortex center. Once the bubbles reach the flow reversal region, the upward moving liquid (in addition to buoyancy) carries them back toward the free interface where they can disengage. On the contrary, a flow field of type 3 is not a desired flow field (Erdal and Shirazi, 2002) since bubbles gathered around the vortex center (in the form of a bubbly filament) will be dragged downward by the liquid, what contributes to more gas carry-under.

At the end of this paragraph, it is interesting to discuss experimental data reported by (Erdal and Shirazi, 2002), comparing the flow fields, respectively, when the swirl is generated using a single or a double tangential inlets. As mentioned above, for a high Reynolds number flow (Re=66,855), when a unique inlet was used, the axial flow was of type 2, while it was of type 3 when the swirl was generated by two diametrically opposed tangential inlets. For a swirl flow at moderate Reynolds number (Re=9285), the mean axial velocity profile admitted a single flow reversal for both situations where the single or the double inlets were used. In this case, the use of two inlets resulted in a wider upward flow region. This means that bubbles can be "captured" more easily, and elevated to the free interface for separation. Also, when two inlets are used, the swirl intensity decays slower than with the single inlet case (Erdal and Shirazi, 2002). These two factors are beneficial and should improve the separation efficiency in the GLCC. However, the double inlet geometry has not been adopted for GLCCs, because as mentioned above, for high flow rates, the associated axial flow field admits a double flow reversal.

Mean Radial Velocity

To the authors' knowledge, no measurements of the radial velocity component in swirl flows in straight pipes have been reported in the literature. In the case of axisymmetric swirl flows,Algifri et al. (1988),Kitoh (1991)andChang and Dhir (1994)calculated the mean radial velocity by using the continuity equation. They found that the mean radial velocity is about two or three orders of magnitude less than the bulk fluid velocity. For a swirl flow generated with a single tangential inlet, numerical simulations showed that the mean radial velocity is of the same order of magnitude than the bulk velocity (Hreiz et al., 2011). The radial velocity component is mainly due to the eccentricity of the vortex center.

Turbulence Kinetic Energy

In swirl flows, the turbulence kinetic energy is strongly coupled with the stabilizing/destabilizing effects of the centrifugal force, which will determine whether the fluctuating components will be absorbed or not by the mean flow. Apart in the vortex core, the swirl presence increases significantly the turbulence level compared to swirl-free axial pipe flows (Algifri et al., 1988,Kitoh, 1991andErdal and Shirazi, 2002). In the vortex core, some investigators reported high turbulence kinetic energy levels (Algifri et al. (1988)andKitoh (1991)who used hot wire anemometers, andErdal and Shirazi (2002) who used LDV) that can even increase while going downstream. However, according to the Rayleigh criterion, the swirl has a stabilizing effect in the forced vortex region: the high turbulence levels reported by these authors are not due to "real turbulence", which is low in the vortex core (Hreiz et al., 2011andKitoh, 1991). In fact, the vortex core undergoes a coherent precession movement (precessing vortex core instability) (see Section4.1), which induces an important periodic variation of the velocity magnitude: the measured high "turbulence levels" in the vortex core are due to this coherent variation of the mean velocity and not to turbulent agitation.

Swirl Decay

Going downstream of the swirl generating device (the tangential slot in the case of the GLCC), the swirl motion decreases due to viscous dissipation. A common way to describe swirl dissipation is to quantify the swirl using dimensionless numbers. However, there is no standard and the swirl number definition may vary between authors.Chang and Dhir (1994)defined the swirl number in a cross-section as:

$$Sn = \frac{\int_0^{2\pi} \int_0^R \rho UWr dr d\theta}{\pi \rho R^2 U_{av}^2}$$

Where ρ is the fluid density, U the mean axial velocity, W the mean tangential velocity, R the pipe radius, U_{av} the fluid bulk velocity, r and ϑ, respectively, the radial and angular position in cylindrical coordinates. They reported that except near the inlet, swirl decays exponentially. For axial distances less than two pipe diameters downstream the inlet, the swirl intensity was nearly constant.

EXPERIMENTAL PROGRAM AND FLOW VISUALIZATIONS

Experimental Facility

A laboratory scale GLCC was designed and built (Fig. 5). The GLCC body is transparent to allow visualizations, and is manufactured in Plexiglas. Its internal diameter, D, is 72mm, and its height is 2.5m. The pipe consists of several sections allowing modification of its height and of the inlet nozzle position. It has several threaded holes to install pressure sensors, or to introduce measurement probes. The lower part of the cyclone is surrounded by a rectangular column filled with water to minimize optical distortions due to the curvature of the GLCC pipe. This is useful for optical or LDV measurements. The liquid vortex height in the GLCC can be controlled through a valve installed on the lower outlet. A centrifugal pump fed by a storage tank provides tap water at a maximum flow rate of 20m³/h. The liquid flow rate is measured by a calibrated rotameter. A centrifugal fan provides air at a maximum relative pressure of 0.8bar, and a maximum flow rate of 550m³/h in standard conditions. Therefore, the GLCC operates almost at atmospheric pressure. Air flow rate and temperature are measured by a mass flow meter (MF50S, Brooks©) based on a thermal measurement technique.

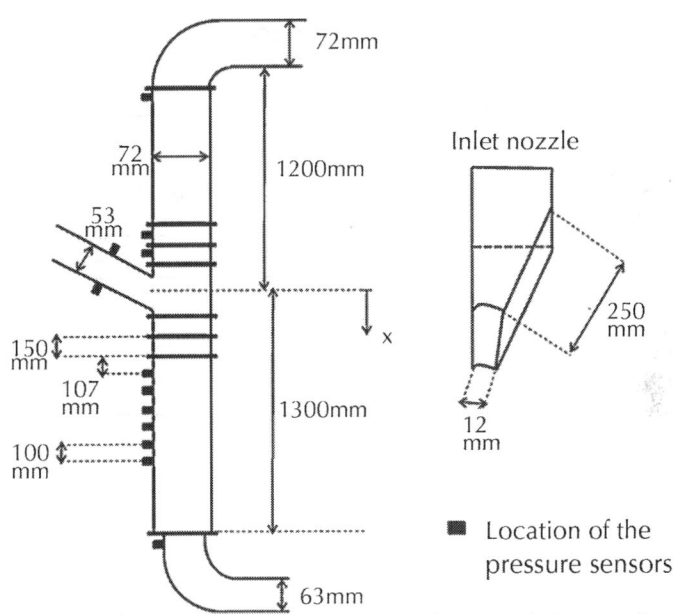

Figure. 5: Main dimensions of the GLCC (not to scale).

Flow rates of both phases are regulated thanks to several valves. Incoming liquid and gas lines are connected to the GLCC inlet channel though aYjunction. The inlet channel has a square cross section as recommended byHreiz et al. (2011). The angle of inclination of the inlet channel is 27° downward, its length is 1.5m, and its section dimensions are 53mm×53mm. The two-phase mixture enters the GLCC body through a converging nozzle. The vertical dimension remains constant along the nozzle (53mm, as the inlet channel height). The nozzle ends up with a horizontal dimension of about 12mm at its intersection with the cyclone body (seeFig. 5).

LDV Measurements

So far, no study has reported velocity measurements data in the GLCC, or in gas–liquid swirl flows comparable to those occurring in the GLCC lower part. Trying to overcome the lacks in experimental data in the literature, local velocity measurements were carried out in the GLCC lower part. A back-scatter 2D LDV (Aerometrics©) operating in the fringe mode and equipped with a RSA-1000L processor was employed to measure the axial and tangential velocities. Laser light is provided by an Ar+laser with a maximal power of 2W.

The laser provides two emission rays, a green one with a wavelength of 514.5nm, and a blue one with a wavelength of 488nm. Each laser beam is split in two synchronous beams with equal intensity.

Directional ambiguity is removed through the use of a 40MHz acoustical modulator Bragg cell. The half-angle of the beams is equal to 2.28°, and the beam pairs are focused using a 250mm focal length lens. In order to reduce the effects of the pipe curvature, the lower part of the cyclone is surrounded by a rectangular column filled with water. This renders the location of the measurement volume for each component (axial and tangential) approximately equal.

Local velocity measurements are conducted for four experimental conditions, whose operating conditions are listed in Table 1. In the first investigated case, the vortex pattern is a deeply excavated vortex flow (Fig. 2c). Compared to single-phase swirl flows, here, the free interface has a considerable effect on the flow. The vortex crown position was fixed at about 21cm below the inlet level. The second investigated case corresponds to an excavated vortex flow (Fig. 2b). Only water is injected in the GLCC.

The vortex crown is also about 21cm below the inlet. The third experimental case differs from the second by the fact that a gas flow rate is also introduced into the separator, what increases the bubble generation rate.

The operating conditions of the fourth experimental case are identical to the second, except that the vortex crown is located at 36cm below the inlet level. Compared to the second investigated case, as swirl is attenuated and gravity effects are enhanced, the corresponding flow pattern is the bubbly vortex flow (Fig. 2a). The pressure at the bottom of the GLCC is measured by a pressure sensor placed at the elbow of the lower inlet (Fig. 5). This information is useful if one would simulate our experiments via Computational Fluid Dynamics (CFD).

Table 1: Parameters of the investigated experimental cases.

Case number	1	2	3	4
Water flow rate, $Ql(m^3/h)$	7.05	3.24	3.24	3.24
Water superficial velocity in the cyclone, $Vs,c,l(m/s)$	0.48	0.22	0.22	0.22
Water Reynolds number, $Rel=\rho lVs,c,lD/\mu l$	34 560	15 840	15 840	15 840
Air flow rate, $Mg(kg/h)$	35	0	110	0
Air superficial velocity in the cyclone, $Vs,c,g(m/s)$	1.99	0	6.25	0
Air Reynolds number, $Reg=\rho gVs,c,gD/\mu g$	17 194	0	54 000	0
Position of the vortex crown below the inlet level	$x=21cm$	$x=21cm$	$x=21cm$	$x=36cm$
Gauge pressure at the GLCC lower outlet (mbar)	154	122	125	100
Vortex flow pattern	Deeply excavated	Excavated (close to the transition to bubbly vortex)	Excavated (close to the transition to bubbly vortex)	Bubbly vortex

To enhance the measurements acquisition rate, the flow is seeded with Iriodine I153 (Merck©) silvery particles. The particles density is about 2.7–2.9g/cm³. Their size distribution ranges between 3 and 110µm, with a mean diameter of 47µm. LDV measurements are performed in the GLCC mid-plane that is perpendicular to the plane ofFig. 5. 49 measurement lines spaced out by 10mm are selected in the measurement plane at a distance going from 320 to 350mm, and from 430 to 870mm below the inlet level (LDV measurements cannot be done betweenx=350mm andx=430mm due to the presence of a flange). At each measurement line, measurements are made at 52 equally spaced locations along the diameter of the GLCC. Considering laser reflections at the walls which make the LDV signal noisy, no measurement can be done at a distance less than 1mm from the walls. At each measurement point, axial velocity, tangential velocity and turbulent quantities are measured during 2min, in order to better take into account slow periodic phenomena in the flow.

RESULTS AND DISCUSSION

The Bubbly Filament

The bubbly filament corresponds to bubbles that gather around the vortex center, permitting to visualize it. When the liquid vortex height is fixed at about 21cm below the nozzle level, the bubbly filament forms if the liquid superficial velocity in the cyclone$Vs,c,l$is superior to 0.03m/s (seeHreiz et al. (2014)for details).

The bubbly filament presents a very complex hydrodynamics, and is subject to different wave types, some of them traveling upstream. This complex hydrodynamics has been characterized by direct flow visualization and by photos taken by a high speed camera. We underline that the bubbly filament behavior is dictated by the vortex core hydrodynamics. It should be kept in mind that the phenomena discussed thereafter would occur with or without the presence of the bubbles (which act as a "tracer", and do not influence too much the vortex core hydrodynamics).

The following observations emerged from the flow visualizations:

The bubbly filament undergoes a coherent precession movement (precessing vortex core instability) as mentioned in Section2.2.4. As reported byHreiz et al. (2011), this coherent movement is responsible of the high "turbulence" levels measured in the core of the flow by different researchers (Algifri et al., 1988,Kitoh, 1991andErdal and Shirazi, 2002). Real turbulence in the forced vortex region is minimal, in accordance with the Rayleigh criterion. In fact, if turbulence was high in the vortex core, bubbles would be dispersed and the bubbly filament would not be able to form correctly. For low liquid flow rates (and thus low swirl intensities), the bubbly filament is fleeting, and often loses its integrity. For high swirl intensities, the stabilizing effect of the swirl motion in the vortex core becomes important.

Thus, the bubbly filament becomes "stable" and "laminar". Fig. 6andFig. 7(see also Video "Bubbly filament hydrodynamics". The video is slowed down 20 times. It represents the same case as inFig. 7) show the bubbly filament in its laminar state (in the first figure, the bubbly filament is quite larger than in the second due to a higher bubble hold-up in the flow). Tracking of relatively large gas bubbles present in the bubbly filament (Fig. 6andFig. 7) reveals that, whenever the bubbly filament becomes laminar, its moves downward (the same direction as the mean flow).

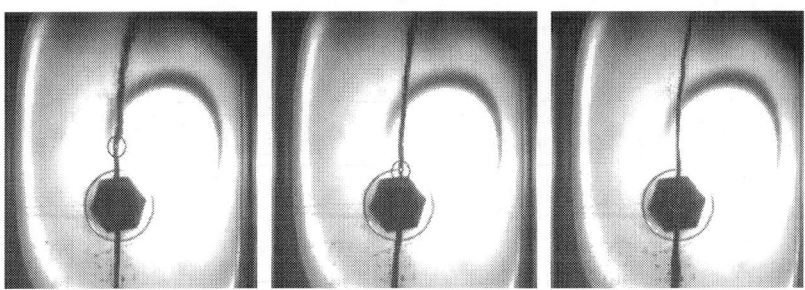

Figure. 6: Laminarization of the bubbly filament.Q_l=6.04m^3/h ($U_{av,l}$=0.412m/s), M_g=55kg/h ($U_{av,g}$=3.13m/s). The vortex crown is atx=21cm. The bubble surrounded by the red circle shows the downward motion of the bubbly filament. The center of the plug in the background is aroundx=100cm. Lightning is done by a small lamp. The time interval between consecutive photos is 10ms. (For interpretation of the references to color in text, the reader is referred to the web version of this article).

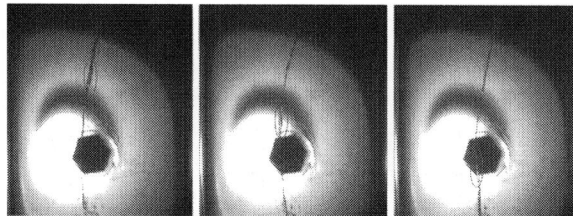

Figure. 7: Laminarization of the bubbly filament. Q_i=9.06m³/h ($U_{av,l}$=0.618m/s), M_g=10kg/h ($U_{av,g}$=0.568m/s). The vortex crown is at x=21cm. The bubble surrounded by the red circle shows the downward motion of the bubbly filament. The center of the plug in the background is around x=100cm. Lightning is done by a small lamp. The time interval between consecutive photos is 10ms. (For interpretation of the references to color in text, the reader is referred to the web version of this article).

When the flow pattern corresponds to an excavated or deeply excavated vortex flow, as the involved swirl intensities are high, laminarization of the bubbly filament takes place often. If air is injected in the cyclone, the flow gets perturbed: the bubbly filament gets laminar less often and breaks off locally more frequently. After getting laminar, the bubbly filament is subjected to a "turbulent puff" wave. A closer look to these turbulence spots indicates that they correspond to a "traveling vortex breakdown" wave, as shown in Fig. 8 (see also Video "Bubbly filament hydrodynamics"). This wave type always induces axially upward velocities at the vortex core.

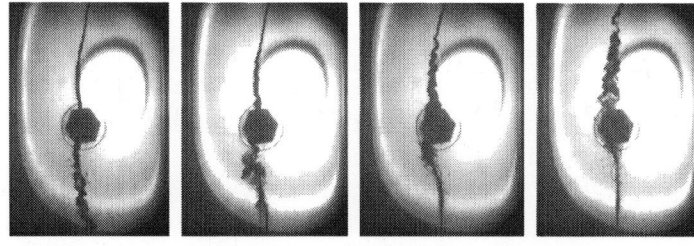

Figure. 8: Turbulent puff of vortex breakdown type crossing the bubbly filament shown in Fig. 6. The time interval between consecutive photos is 15ms.

Whenever it forms (Vs,c,l>0.03m/s), the bubbly filament stretches down to the lower outlet. This is observed even when the bubble density is extremely low, for example when the flow pattern is the deeply excavated vortex. In these cases also, the bubbly filament contains important amounts of bubbles despite the fact that bubble density is extremely low in the surrounding liquid (Fig. 6,Fig. 7andFig. 8). This indicates that under these operating conditions, bubbles are not conveyed toward the bubbly filament by radial transport, but are convected axially downward. Moreover, bubbles coalescence can take place in the bubbly filament: bubbles much larger than in the surrounding liquid are observed in the bubbly filament (Fig. 6andFig. 7). This means that in the bubbly filament, bubbles remain in contact for important durations, i.e. their residence time in the vortex core is important (what allows their coalescence). This indicates that liquid drag force is opposite to buoyancy.

These different observations suggest that, for high liquid flow rates, the mean axial velocity near the vortex center points downwards. Thus, the hypothesis admitted thus far that the mean axial velocity in the GLCC is described by the profile shown inFig. 4b does not hold for the high swirl intensity cases: bubbles in the bubbly filament are not brought back toward the free interface and separate, but are transported toward the GLCC lower outlet and cause GCU. This means that for high liquid flow rates, the mean axial flow admits a double flow reversal (Fig. 4c). This conclusion is confirmed quantitatively by LDV measurements in the GLCC lower part (Section4.3.2).

For deeply excavated vortex flows, a low-frequency coherent phenomenon, the vortex oscillation, interferes with the mean velocity field. The vortex crown (i.e. the upper part of the free interface, seeFig. 1) is wide and oscillates as a block. Its oscillations seem related to the instantaneous fluctuations in the feeding flow rate, as the flow pattern at the inlet corresponds to a stratified wavy or a slug flow regime (Hreiz et al., 2014). The nose of the vortex (i.e. the bottom of the free interface, seeFig. 1) is generally sharp and narrow, with a size similar to that of the bubbly filament section. It oscillates at low frequency, penetrating/withdrawing itself rapidly from the area occupied by the bubbly filament (which corresponds to a low pressure region) (Fig. 9. See also Video "Vortex nose oscillations". The video is slowed down 20 times. It represents the same case as inFig. 9). If the air flow rate is increased, the flow at the inlet becomes more disturbed or switches to a slug flow pattern. In this case, the nose of the vortex tends to round up and the amplitude of its oscillations is attenuated.

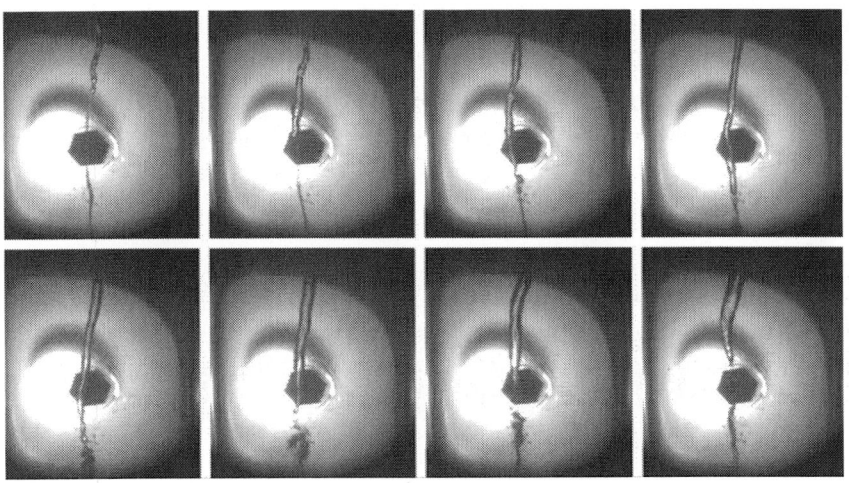

Figure. 9: Oscillation of the vortex nose.Q_j=9.06m³/h ($U_{av,l}$=0.618m/s),M_g=10kg/h ($U_{av,g}$=0.568m/s). The vortex crown is located atx=21cm. The center of the plug in the background is aroundx=100cm. The time interval between consecutive photos is 10ms. Note that the vortex withdraws when it encounters a vortex breakdown wave, which corresponds to high pressure spot.

Localization of the Vortex and of the Bubbly Filament in the First Investigated Experimental Case

For the first investigated experimental case (seeTable 1), as the vortex flow pattern is deeply excavated, the position of the free interface has a considerable effect on the hydrodynamics behavior of the flow. For this reason, before reporting local velocity measurements in the GLCC, it is important to localize the mean position of the vortex and of the bubbly filament in the LDV measurement plane. Such information will also be useful for future comparison with CFD simulations.

Photos of the flow in the GLCC lower part have been taken by a high speed camera from two angle shots (Fig. 10a), and at several axial positions. The first angle shot, to which we will refer as "side view", corresponds to the view inFig. 5. The second angle shot is the "front view", where the observer is in front of the inlet nozzle. The pixel size in the different series of photos is about 100μm×100μm. Photos are processed using the ImageJ software, which allows calculating the position of the vortex and of the bubbly filament, as observed from these angle shots. To calculate the void fraction, it is considered that the bubbly filament is constituted from gas only. The void fraction due to individual bubbles (not gathered in the bubbly filament) is neglected during image processing. The hypothesis is quite acceptable considering the very low amount of bubbles in the vortex and their small size.

Results of image processing (temporal averages) are shown inFig. 10b and c. Sections filled only with black correspond to positions that could not be processed because of the presence of a flange or a plug in the background, or to zones that were badly enlightened in the photos.Fig. 10b and c reveals a helical aspect of the zone occupied by the vortex and the bubbly filament. The warping of the vortex is due to the non-axisymmetric conditions of swirl generation.

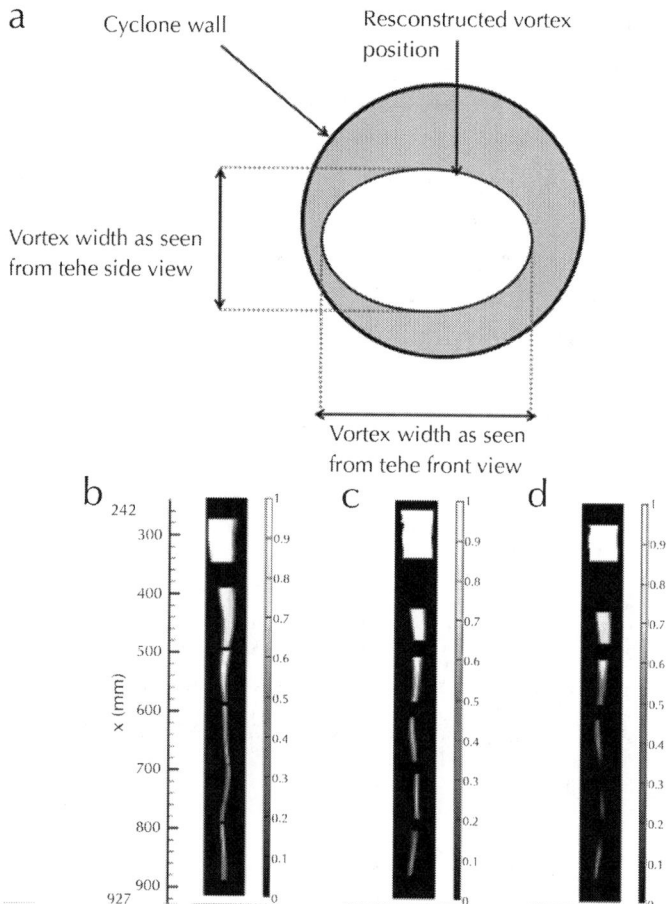

Figure. 10: Mean void fraction in the GLCC lower part for the first inves-tigated experimental case. (a) Mean void fraction as seen from the side view. (b) Mean void fraction as seen from the front view. (c) Estimated mean void fraction in the LDV measurement plane. (d) "Reconstruction" of the vortex (and bubbly filament) mean position.

Void Fraction in the LDV Measurements Plane

In order to estimate the mean void fraction in the LDV measurements plane, we considered that the vortex (and the bubbly filament) cross-section has an ellipsoïdal shape. The lengths of the major and minor axis of the ellipse are taken equal to the mean width of the vortex, respectively, as seen from the side and front views. Considering this assumption, it is possible to "reconstruct" the mean position of the vortex and of the bubbly filament in 3D (Fig. 10a). The mean void fraction at a given position is considered equal to the product of the corresponding mean void fractions as seen from the side and front views. The estimated mean void fraction contour in the LDV measurements plane is shown inFig. 10d.

Variation of the Void Fraction in the Axial Direction

As the mean position of the vortex and of the bubbly filament has been estimated, it is possible now to deduce the mean void fraction in the GLCC cross-sections.Fig. 11shows the variation of the mean void fraction in the GLCC cross-section with respect to the axial distance from the inlet (note that the graph is plotted in semi-logarithmic coordinates). Several observations can be made:

- Close to the vortex crown, the mean void fraction is nearly constant.

- For a distance from the inlet level betweenx=40cm andx=60cm, the mean void fraction decreases exponentially with respect to the axial coordinate. However, two different decrease coefficients are obtained, respectively, forxbetween 40 and 50cm, and forxbetween 50 and 60cm. We note that in the first zone, the vortex is always present, whereas the vortex nose penetrates and withdraws from the second zone because of its axial oscillations. The difference between the two decreasing coefficients can be due to this phenomenon or to the numerical processing errors and to the approximations used.

- In the zone where only the bubbly filament is present ($x>60$cm), the mean void fraction remains nearly constant.

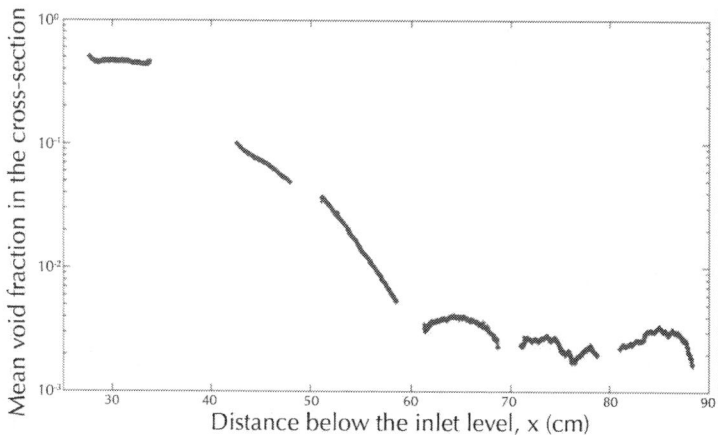

Figure. 11: Mean void fraction in the GLCC cross-sections versus the axial distance from the inlet level (experimental investigated case 1).

It can be noted that in the cross-sections where the vortex is present, the profile of the mean void fraction presents the same characteristics as the profile of the swirl intensity in single-phase swirl flows (Section2.2.5): it is nearly constant in the region close to the inlet level, and then decays in an exponential manner. Thus, it seems that the excavated depth of the vortex is strongly correlated with the swirl intensity in the flow.

LDV Results

In Section3.2, it has been mentioned that the laser of our LDV setup supplies a maximum power of 2W. However, we observed that the laser had some problems. In fact, the generated power falls during the manipulations, and the laser must be adjusted to find its initial intensity. This loss of intensity leads to a reduction in the acquisition frequency of the LDV. The intensity of the Doppler bursts from the seeding particles (and from the small bubbles) decreases, so, when collected by the photodetector, it can be considered as noise and is not processed. For this reason, in the next paragraphs where our results are reported, the reader will remark a lack of measurements at some positions in the LDV measurements plane.

When the vortex is excavated or deeply excavated, it is not possible to perform velocity measurements behind it. One or both laser beams (of a pair) are deviated by the vortex interface (reflected and refracted), and so, the laser beams do not intersect to form a measurement volume. This problem is also encountered with the bubbly filament. However, as the bubbly filament undergoes a precessing movement, in some cases, laser beams can intersect and form a measurement volume momentarily behind it, and measurements can be performed. Due to this constraint, measurements could not been conducted behind the vortex and at some positions behind the bubbly filament.

Another measurement complication arises from the presence of numerous reflective interfaces (vortex interface, bubbly filament, cyclone walls, etc.). Laser reflections (and multi-reflections) from these interfaces keep a high intensity. Thus, if they hit the photodetector, due to their important intensity, they get processed (instead of being treated as noise). However, as these interfaces are slow-moving or stationary, velocities arising from reflections are in the form of a distribution centered not far from 0m/s (Fig. 12). At each measurement point, it is necessary to remove the velocity spectrum originating from reflected laser light, so the mean velocities and turbulent quantities can be calculated. However, when the real velocity distribution is centered around 0m/s, the two velocity spectra overlap, and it is no more possible to separate them. For this reason, we defined degrees of reliability of the measurements:

- A degree 3 corresponds to measurements from which the reflections spectrum has been completely removed.
- A degree 2 corresponds to measurements from which the reflections spectrum has been only partially removed. The calculated mean velocity is still enough accurate. The calculated turbulent quantities must be regarded with caution.
- A degree 1 corresponds to strongly altered measures from which it has not been possible to remove the reflections spectra. They are shown for illustrative purposes only.

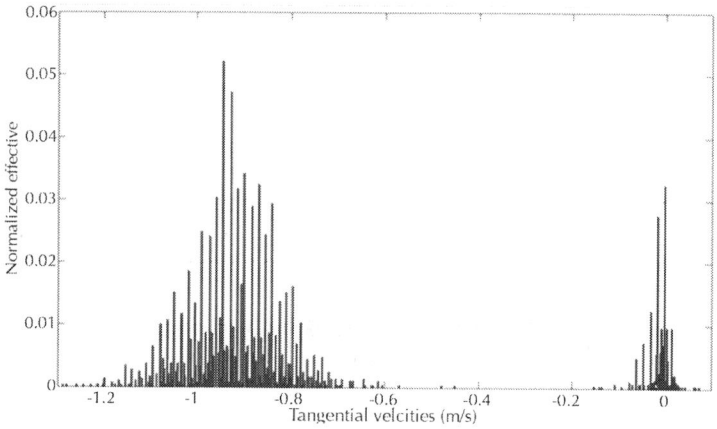

Figure. 12: Measured velocities distribution: the distribution at the left corresponds to "real" liquid velocities, whereas the distribution at the right is due to the contribution of reflected laser beams.

Mean Tangential Velocity

LDV measurements have been used to create color contour plots of the mean axial and tangential velocities, and of the root of the corresponding diagonal components of the Reynolds stress tensor. Contour plots of the mean tangential velocity field (normalized withVs,c,l) for the different experimental cases investigated (seeTable 1) are shown inFig. 13. The last contour plot is shorter than the previous ones because the vortex level was set lower, so no measurements are performed for$x$between 320 and 350mm for this case. Negative velocities represent tangential velocities pointing out of the page, and positive ones represent flow into the page. Mean tangential velocity profiles for the different experimental cases at distances$x$=500, 600 and 750mm below the inlet level are shown inFig. 14. More velocity profiles data can be found inHreiz (2011).

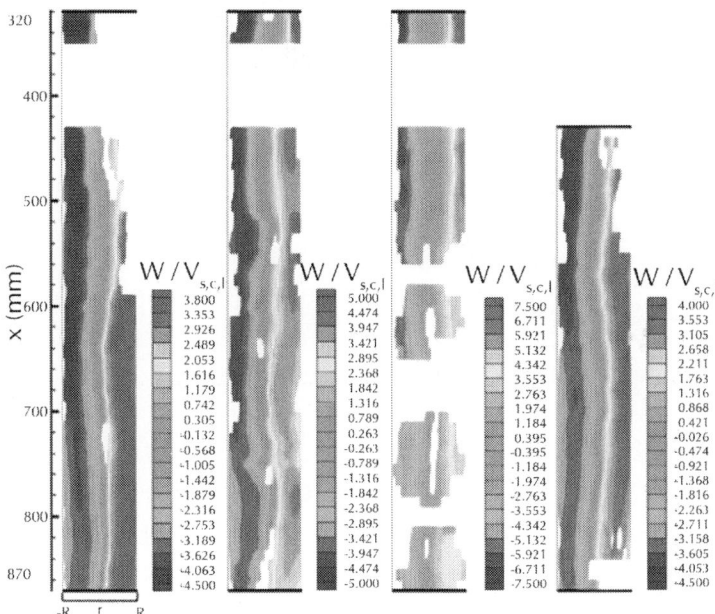

Figure. 13: Contour plots of the mean tangential velocity: from left to right, experimental cases from 1 to 4, respectively.

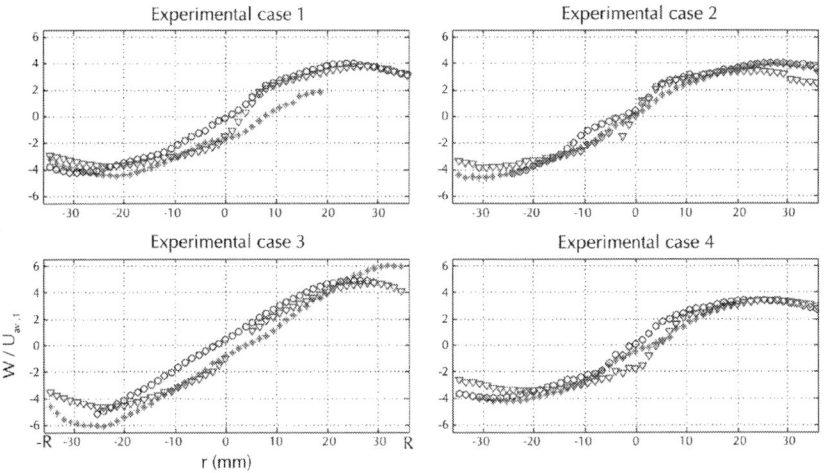

Figure. 14: Profiles of the mean tangential velocity for the different investigated experimental cases. Red stars, black circles and blue triangles

represent measurements atx=500, 600 and 750mm, respectively. (For interpretation of the references to color in text, the reader is referred to the web version of this article.)

Fig. 14shows that the tangential velocity field has a similar structure than in single-phase swirl flow (Section2.2.1). The central region corresponds to a forced vortex, while the surrounding zone corresponds to a free vortex flow (with a transition region between the free and forced vortex). No velocity data are reported in the thin near wall layer because LDV measurements cannot be performed at a distance less than 1mm from the wall. The mean tangential velocities are high in the free vortex region, and decrease toward the center. The maximum tangential velocity magnitude decreases in the downward axial direction due to swirl decay. Location of zero mean tangential velocity has a helical shape due to vortex warping.

The mean tangential velocities are higher in experimental case 2 than in case 4, due to higher swirl intensity. Tangential velocities magnitude is higher in case 3 than in case 2 as the gas flow accelerates the liquid rotation. Another reason may be that the presence of a gas flow leads to higher liquid velocities at the nozzle outlet (by reducing the effective flow section of the liquid phase). Results show that the injection of an air flow rate in the cyclone leads to an increase in the extent of the forced vortex region (Fig. 14).

Mean Axial Velocity

Fig. 15shows the contour maps of the mean axial velocity field (normalized withVs,c,l) for the four investigated experimental cases. Negative velocities represent axial velocities pointing in the same direction than the mean flow, i.e. downward, and positive ones represent reverse flow. Mean axial velocity profiles for the different experimental cases at distances$x$=500, 600 and 750mm below the inlet level are shown inFig. 16. More velocity profiles data can be found inHreiz (2011).

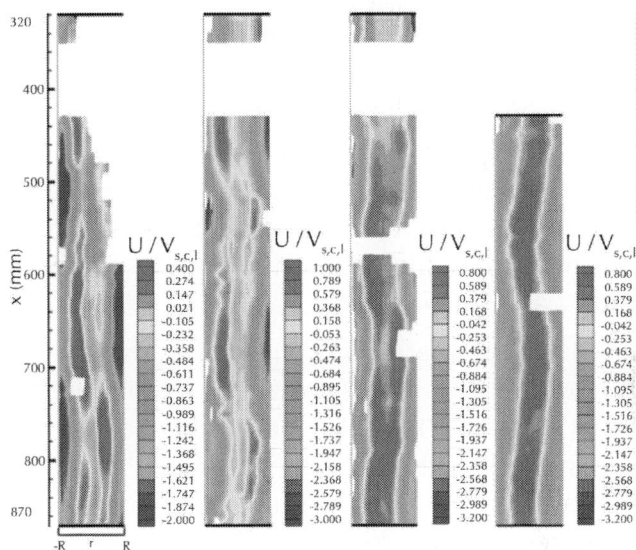

Figure 15: Contour plots of the mean axial velocity: from left to right, experimental cases from 1 to 4, respectively.

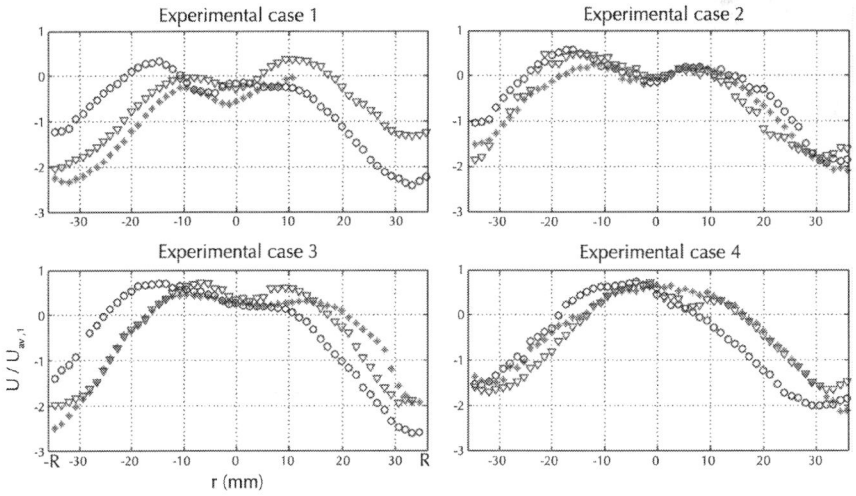

Figure. 16: Profiles of the mean axial velocity for the different investigated experimental cases. Red stars, black circles and blue triangles represent measurements atx=500, 600 and 750mm, respectively. (For interpretation

of the references to color in text, the reader is referred to the web version of this article.)

The contour maps reveal details of the flow behavior. In experimental cases 3 and 4, the mean axial flow has a single flow reversal. In experimental cases 1 and 2, the mean axial flow presents a double flow reversal, with downward velocities around the zone occupied by the bubbly filament (and near the vortex interface in experimental case 1, seeFig. 10d).

The first question that arises further to these results is the origin of the double flow reversal regime. According to our visual observations (Section4.1), near the vortex center, downward velocities are induced by the laminarization of the vortex core (Fig. 6andFig. 7). Upward velocities are due to the adverse axial pressure gradient that arises due to swirl motion. In experimental cases 1 and 2, the involved swirl intensity is high, what leads to the excavation of the vortex. High swirl intensities promote the stability of the flow in the forced vortex region (Rayleigh criterion). As a consequence, the vortex core laminarization is favored, and gets more important than the adverse pressure gradient effect: this results in downward mean velocities around the vortex center. In experimental case 4, due to lower swirl intensity, the vortex core laminarization takes place less often. Thus, its hydrodynamics is dominated by the adverse pressure gradient: this leads to upward mean velocities around the vortex center. Experimental case 3 involves a high swirl intensity, however, the corresponding mean axial flow presents a single flow reversal. In fact, the air flow disturbs and destabilizes the flow in the liquid vortex. Photos taken by high speed camera reveal that the bubbly filament gets agitated, often breaks off locally, and gets laminar less frequently than in experimental case 2. As the laminarization phenomenon is disadvantaged, the adverse pressure gradient dominates the flow dynamics at the vortex core, and leads to a single flow reversal mean axial velocity field.

According to the literature on single-phase swirl flows (Section2.2.2), the double flow reversal regime has only been encountered when the swirl is generated by multiple tangential inlets. In fact, as reported byErdal and Shirazi (2002), when multiple tangential inlets are used, decay of swirl is lower than when a unique inlet is employed (so the swirl intensity is overall higher). This favors the vortex core laminarization what results in the double flow reversal regime. In our experiments, this profile type has been encountered with a single tangential injection, most likely due to the severe convergence of the inlet nozzle, which imparts high swirl intensities to the flow. The presence of a free interface may also influence the onset of this regime; however, its role seems secondary since this flow structure has been encountered in single-phase swirl flows.

Our results provide a better understanding of the origin of the double flow reversal regime, by linking it to the hydrodynamics of the vortex core (revealed thanks to the bubbly filament). However, it has not been possible to explain why the vortex core laminarization induces downward velocities. The phenomenon seems quite complex, and could be related to other phenomena taking place at the vortex core, as the precessing vortex core instability.

Concerning the GLCC separation efficiency, the axial velocity measurements lead to interesting conclusions. In fact, as reported byHreiz et al. (2014), the GCU (Gas Carry-Under) level is lower when the vortex flow pattern is deeply excavated, than when it corresponds to a bubbly vortex flow. According to the LDV results and to the discussion made in this paragraph, the more likely mean axial velocity profile in the GLCC lower part is of type 3 (i.e. double flow reversal, seeFig. 4c) for high swirl intensity flows (case of the deeply excavated flow pattern), and of type 2 (i.e. single flow rever-sal, seeFig. 4b) for low to intermediate liquid flow rates (case of the bubbly vortex pattern). Given these observations, it is concluded that contrary to the commonly assumed hypothesis (Erdal and Shi-razi, 2002), a flow regime with a double flow reversal is beneficial for GLCC performance.

In fact, high swirl intensity flows for which the axial flow regime is of type 3 are associated with low bubble generation rates in the GLCC lower part, and in consequence, to low bubble hold-up in the vortex (seeHreiz et al. (2014)for details). Bubbles are gathered in the bubbly filament due to high centrifugal effects, and will be carried under due to the mean downward velocities around the vortex center: separation efficiency in terms of percentage of bubbles that is separated is low. However, as a low quantity of bubbles is found in the vortex, the resulting GCU is low: separation efficiency in terms of amount of gas carried under is very good.

On the other hand, an axial flow regime of type 2 is associated with relatively low/moderate swirl intensity flows. For this range of liquid flow rates, high bubble dispersion can occur in the GLCC lower part. Bubbles that reach the flow reversal region are convected to the free interface and separate. However, the centrifugal forces are not sufficient to transport all the bubbles toward the vortex center, so part of the bubbles cannot be removed in time, and exits with the liquid stream. As bubble hold-up can be important, the fraction of bubbles that are carried under results in higher GCU levels than when the axial flow structure is of type 3. Further to the results presented in this paragraph, we propose the use of multiple tangential inlets for a better efficiency of the GLCC. This inlets configuration has already proven their efficiency with other cyclone separators (Tue-Nenu and Yoshida, 2009). The use of multiple inlets promotes an axisymmetric flow field in the GLCC and reduces the vortex warping (due to smoother swirl generation conditions): this will reduce liquid "short circuiting" toward the upper outlet, and expand the GLCC operational limit with respect to liquid carry-over (Hreiz et al., 2014). Also, when multiple inlets are used, swirl decays slower than with the one inlet configuration (Erdal and Shirazi, 2002), which improves the GLCC separation efficiency.

Turbulent Quantities

The root-mean-square tangential velocity (*wrms*) and axial velocity (*urms*) fluctuations (normalized with $V_{s,c,l}$) in the measurements plane are contour plotted in Fig. 17 and Fig. 18, respectively. *wrms* and *urms* profiles for the different experimental cases at distances $x=500$, 600 and 750mm below the inlet level are shown in Fig. 19 and Fig. 20, respectively.

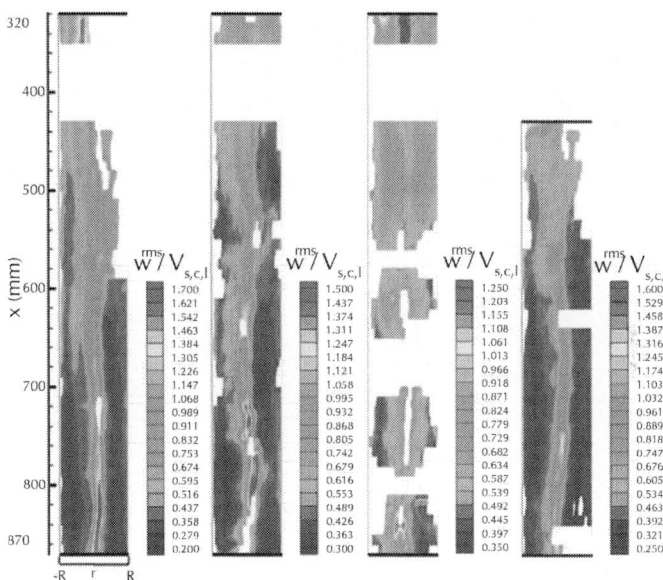

Figure. 17: Contour plots of the root mean square tangential velocity w^{rms}: from left to right, experimental cases from 1 to 4, respectively.

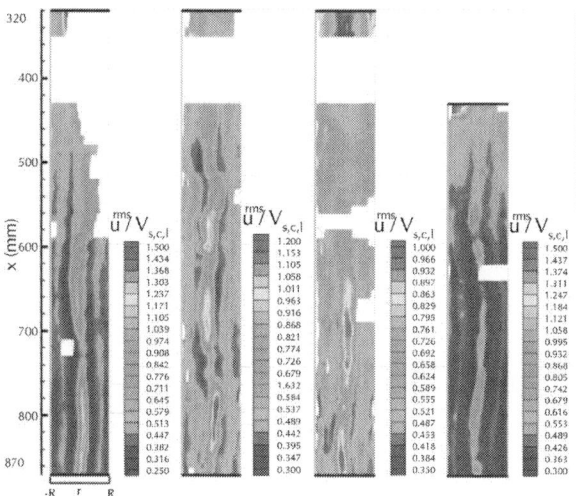

Figure. 18: Contour plots of the root mean square axial velocity u^{rms}: from left to right, experimental cases from 1 to 4, respectively.

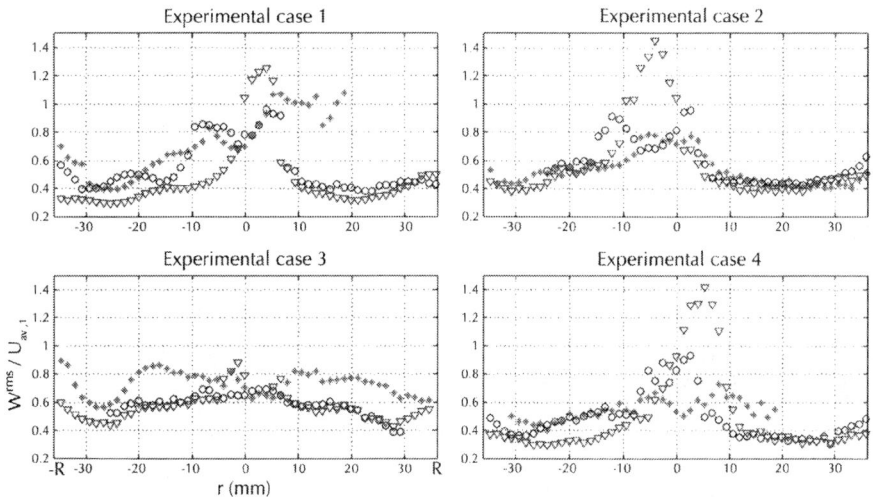

Figure. 19: Profiles of w^{rms} for the different investigated experimental cases. Red stars, black circles and blue triangles represent measurements at $x=500$, 600 and 750mm, respectively. (For interpretation of the references to color in text, the reader is referred to the web version of this article).

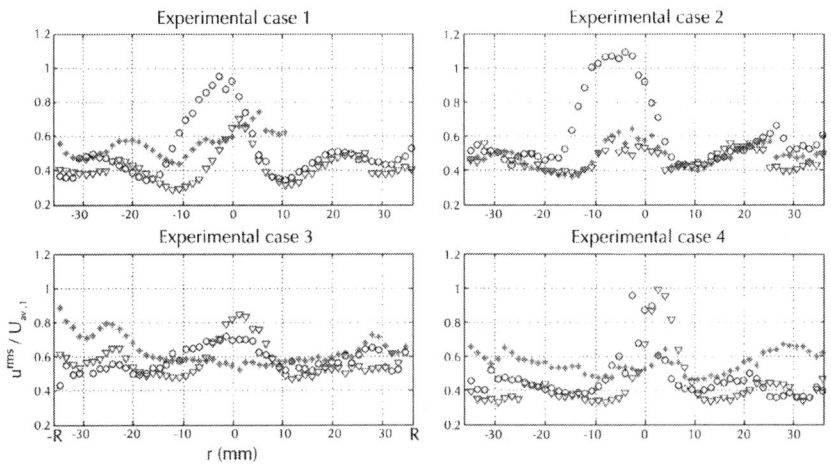

Figure. 20: Profiles of u^{rms} for the different investigated experimental cases. Red stars, black circles and blue triangles represent measurements at $x=500$, 600 and 750mm, respectively. (For interpretation of the

references to color in text, the reader is referred to the web version of this article).

The contours show very high turbulence levels in the annular region near the top, which can be due to the high velocity inlet jet. This important turbulence near the free interface probably contributes to the engulfment of bubbles in the vortex. Turbulence in the annular region decays downward axially.

In the vortex core, very important "turbulence" levels are encountered. As noted in Section2.2.4, and in accordance with the Rayleigh criterion, these important fluctuations are mainly due to non erratic phenomena (precession of the vortex core, oscillation of the vortex nose) and not to turbulent agitation. If turbulence were high around the vortex center, bubbles would be dispersed and the bubbly filament would not be able to form correctly. The highest fluctuation levels near the vortex center inFig. 17andFig. 18correspond to the locations where the bubbly filament intersects the measurements plane, since these velocity fluctuations are mainly due to the precession of the vortex core. As already reported byErdal and Shirazi (2002), fluctuations intensity in the vortex core can increase while going downstream.

Reliability of the Velocity Measurements

Fig. 21a and b shows, respectively, the measurements reliability (evoked in Section4.3) of the axial and tangential velocities. The measurements reliability is lower in zones where the velocity distribution is centered around 0m/s, because "measurements" due to laser reflections (i.e. noise) could not be fully removed from the acquired data.

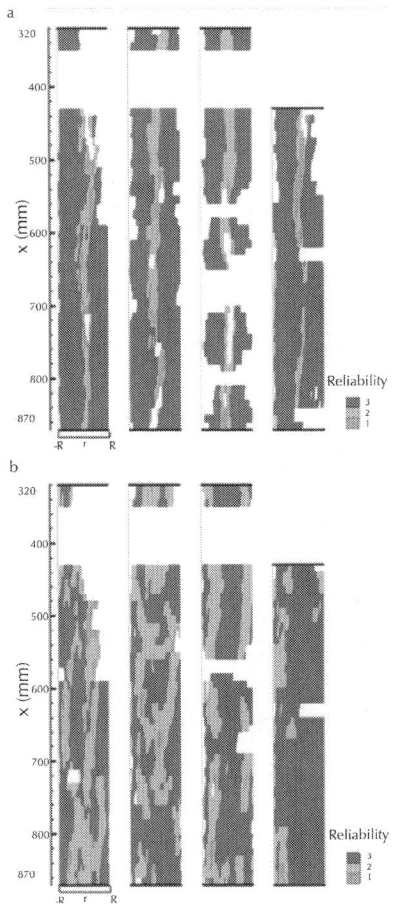

Figure. 21: Contour plots of the reliability of the velocity measurements. (a) Tangential velocity measurements: from left to right, experimental cases from 1 to 4, respectively. (b) Axial velocity measurements: from left to right, experimental cases from 1 to 4, respectively.

However, despite the fact that at some positions measurements are not very reliable, results are globally very good, given that the details of the flow behavior are clearly revealed. It should be noted that velocity measurement in such flows is quite a difficult task, owing their 3D nature and the broad range of involved interfacial scales. To the authors' knowledge, it is the first time that local velocity measurements are conducted in such gas–liquid swirl flows. Some measurements techniques as the hot-wire anemometer probably permit less noisy measurements than optical techniques, but at the expense of non-intrusiveness.

Subsequent Experimental Tests

Insertion of a Solid Rod in the Center of the Cyclone

In an attempt to further improve the bubbles separation efficiency for high liquid flow rates, a new series of experiments has been conducted. A metal rod of 8mm diameter has been inserted in the center of the GLCC lower part: the goal is to prevent the formation of a zone with downward mean velocities around the vortex center. Thus, bubbles gathered in the vortex core will be conveyed back by the upward moving liquid toward the free interface and will be separated.

Flow visualizations revealed that, indeed, even for high liquid flow rates, bubbles gathered around the solid rod are transported upward by the liquid. However, the presence of the rod leads to relatively important swirl dissipation. Both bubble generation rate and bubble hold-up in the vortex increase. Part of the bubbles does not reach the flow reversal region (because of the decrease of the centrifugal forces) and exits with the liquid stream. The GCU is more important than with experiments without the solid rod. Although this method was unsuitable with the laboratory scale GLCC, it may be beneficial for large GLCC prototypes, where swirl attenuation due to the rod may be negligible compared to the total angular momentum of the flow.

Utilization of a Vortex Breaker

Swirl flows are associated with high head losses compared to swirl-free flows. To reduce pumping requirements, in industry, it is quite common to insert a vortex breaker in the pipes downstream of the cyclone to dissipate the swirl motion.

A last series of experiments was performed while inserting a cross vortex breaker (two perpendicular blades) in the bend of the lower outlet. Flow visualizations showed that the vortex breaker presence dramatically decreased the swirl intensity in the cyclone. The deeply excavated vortex regime was not reached even under high liquid flow rates. The separation efficiency was severely altered due to low centrifugal effects. Thus, results demonstrate that the hydrodynamics in the cyclone is quite sensitive to downstream conditions: vortex breakers should not be installed too close from the GLCC outlets.

CONCLUSIONS AND RECOMMEN-DATIONS

The hydrodynamics in the GLCC lower part has been characterized by flow visualizations. Bubbles in the vortex act as a tracer and allow a visualization of the vortex core. The conclusions emerging from these visual observations have been confirmed by LDV measurements. The results show that the bubbly filament presents a very complex hydrodynamics, characterized by an alternation between a laminar and a turbulent state. The laminar regime is associated with downward velocities near the vortex center. Turbulent puffs and the adverse pressure gradient in the vortex core induce upward velocities near the vortex center. High swirl intensities stabilize the flow in the vortex core, thus the bubbly filament get laminar very often: the corresponding mean axial velocity profile presents a double flow reversal, with downward flow around the vortex center, due to the vortex core laminarization. When the swirl intensity in the flow is low, or if the flow in the vortex is disturbed by the injection of an air flow in the cyclone, bubbly filament laminarization occurs less frequently: the corresponding mean axial velocity profile presents a single flow reversal. The present results give a first understanding about the origin of the double axial flow reversal in swirl flows, linking this flow structure to the vortex core laminarization phenomenon. Such a flow structure can have serious impacts on processes employing swirl flows, as cyclone separators or heat exchangers. It also complicates CFD simulations of swirl flows, as turbulence models generally fail to predict flow laminarization (Hreiz et al., 2011).

Concerning the GLCC separation efficiency, it is shown that, contrary to the commonly assumed hypothesis, the double flow reversal regime is associated with good performance of the GLCC. In fact, albeit bubbles separation efficiency is poor, low GCU levels occur, as this regime is associated with low bubble hold-up in the vortex.

Finally, further to the results of this paper, we propose the use of multiple tangential inlets to improve separation efficiency in GLCCs. Such inlet configuration leads to lower swirl intensity decay than the unique inlet configuration. It also engenders a more axisymmetric flow, which would improve the GLCC performance with respect to liquid carry-over

REFERENCES

1.   Algifri, A., Bhardwaj, C., Rao, Y., 1988. Turbulence measurements in decaying swirl flow in a pipe. Appl. Sci. Res. 45, 233–235.

2.   Chang, F., Dhir, V.K., 1994. Turbulent flow field in tangentially injected swirl flows in tubes. Int. J. Heat Fluid Flow 15, 346–356.

3.   Erdal, F., Shirazi, S., 2002. Effect of inlet configuration on flow behavior in a cylindrical cyclone separator. In: ASME Engineering and Technology Conference on Energy.

4.   Fokeer, S., Lowndes, I., Kingman, S., 2009. An experimental investigation of pneumatic swirl flow induced by a three lobed helical pipe. Int. J. Heat Fluid Flow 30, 369–379.

5.   Guo, Z., Dhir, V.K., 1990. Flow reversal in injection induced

swirl flow. In: ASME Single and Multiphase Convective Heat Transfer, 145, pp. 23–30.

6.   Hreiz, R., (PhD dissertation) 2011. Etude expérimentale et numérique de séparateurs gaz-liquide cylindriques de type cyclone. Institut National Polytechnique de Lorraine (in French).

7.   Hreiz, R., Gentric, C., Midoux, N., 2011. Numerical investigation of swirling flow in cylindrical cyclones. Chem. Eng. Res. Des. 89, 2521–2539.

8.   Hreiz, R., Lainé, R., Wu, J., Lemaitre, C., Gentric, C., Fünfschilling, D., 2014. On the effect of the nozzle design on the performances of gas–liquid cylindrical cyclone separators. Int. J. Multiphase Flow 58, 15–26.

9.   Kitoh, O., 1991. Experimental study of turbulent swirling flow in a straight pipe. J. Fluid Mech. 225, 445–479.

10. Kouba, G.E., Shoham, O., Shirazi, S., 1995. Design and performance of gas liquid cylindrical cyclone separators. In: Proceedings of the BHR Group 7th International Meeting on Multiphase Flow, pp. 307–327.

11. Mantilla, I., (Master's thesis) 1998. Bubble trajectory analysis in gas–liquid cylindrical cyclone separators. The University of Tulsa.

12. Martemaniov, S., Okulov, V., 2004. On heat transfer enhancement in swirl pipe flows. Int. J. Heat Mass Transfer 47, 2379–2393.

13. Nissan, A., Bresan, V., 1961. Swirling flow in cylinders. AIChE J. 7, 543–547.

14. Rosa, E.S., Franc¸a, F.A., Ribeiro, G.S., 2001. The cyclone gas–liquid separator: operation and mechanistic modeling. J. Pet. Sci. Eng. 32, 87–101.

15. Shoham, O., Kouba, G.E., 1998. State of the art of gas/liquid cylindrical-cyclone compact-separator technology. J. Pet. Technol., 50.

16. Tue-Nenu, R.K., Yoshida, H., 2009. Comparison of separation

performance between single and two inlets hydrocyclones. Adv. Powder Technol. 20, 195–202.

17. Wang, S., (PhD dissertation) 2000. Dynamic simulation, experimental investigation and control system design of gas–liquid cylindrical cyclone separators. The University of Tulsa.

18. Wegner, B., Maltsev, A., Schneider, C., Sadiki, A., Dreizler, A., Janicka, J., 2004. Assessment of unsteady RANS in predicting swirl flow instability based on LES and experiments. Int. J. Heat Fluid Flow 25, 528–536.

19. Yapici, S., Yazici, G., Ozmetin, C., Ersahan, H., 1997. Mass transfer to local electrodes at wall and wall friction factor in decaying turbulent swirl flow. Int. J. Heat Mass Transfer 40, 2775–2783

Chapter 7

A Novel Centrifugal Gas–Liquid Separator for Catching Intermittent Flows

M. Creutz[a,] and D. Mewes[a]

[a]Institute of Process Engineering, Hannover University, Callinstr. 36, 30167, Hannover, Germany

ABSTRACT

For separating transient two phase gas–liquid flows, a tool is developed that is based on a rotational pump impeller with axial gas removal. Theoretical and experimental investigations are presented concerning the transient behaviour of the separator and the quality of the separation. The basics of the hydrodynamic layout of the separator are presented for steady state and transient conditions.

INTRODUCTION

In various applications two phase gas–liquid flows have to be separated. For that purpose, static gravity separators are often used. For transient slug flow and high pressures at the separator inlet, these are large and costly tools (Hollenberg et al., 1995). For this reason intensive effort has been undertaken to develop separation techniques that use centrifugal forces in order to reduce their dimensions. Kouba et al. (1995) and Nebrensky et al. (1980) developed cyclone separators for the off-shore industry. Muschelknautz and Mayinger (1990) use rotating impellers for separating the liquid–gas mixture of a blowdown process in chemical reactors.

A new concept for transient gas–liquid separation is based on a rotary pump with gas-separation inside the impeller. The separator is shown schematically in Fig. 1. It consists of a stationary casing and a rotating shaft with an impeller. In contrast to the impeller of a centrifugal pump, this impeller is open on both sides. The liquid is forced through the impeller into a ring chamber by means of centrifugal forces and the pressure is increased. The gas penetrates the impeller in the axial direction and leaves the separator. If the flow that enters the separator is intermittent, the liquid flow rate has to be damped in order to obtain a steady liquid outflow with low pressure fluctuations behind the impeller. The liquid flow fluctuations are moderated using a variable positioning of the interface between gas and liquid inside the impeller. A further separation of solid particles from the liquid is additionally obtained through a centrifuge mounted below the impeller. The solid–liquid separation is described by Creutz (1998) and not discussed in the current paper.

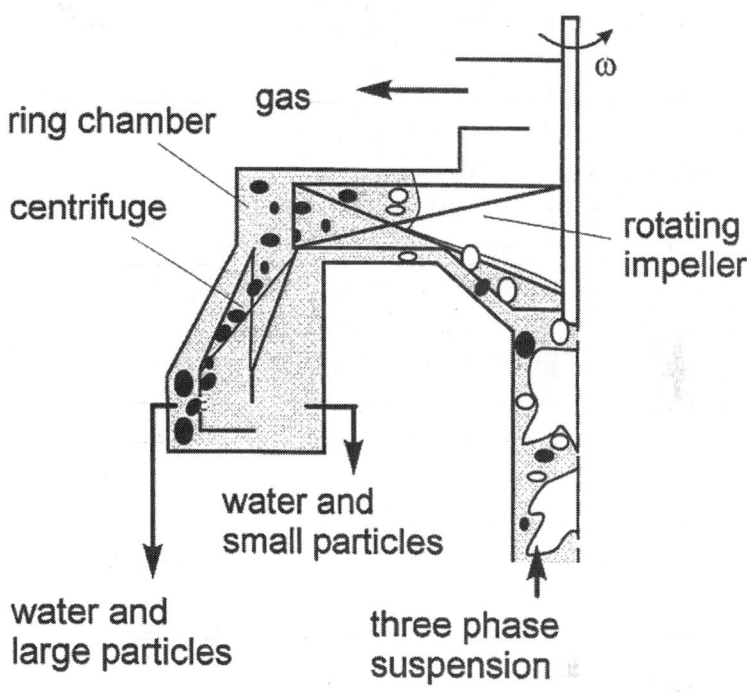

Figure 1: Schematic of the separator.

The two phase separator can be considered as a combination of a slug catcher, a separator and a pump in the process and in the off-shore industry. The three phase separator including the solid–liquid centrifuge is applied in underwater cutting processes where gaseous and solid contaminants are emitted, removed together with water and separated. These contaminants reduce the optic transparency of water. They have to be avoided in processes controlled by systems based on visual observation, especially in cutting processes used for the demolition of nuclear power plants. In the latter case, the contaminants have to be separated under extremely space-limited conditions. The physical basics for the layout of the separator are deduced in the current paper for steady state and dynamic operating conditions. Experimental and theoretical investigations on the dynamic behaviour and on the quality of the gas separation are also presented.

SEPARATOR LAYOUT AND EXPERIMENTAL FACILITY

The technical drawing of the separator manufactured from stainless steel is shown in Fig. 2. The inlet pressure and the back-pressure may vary between 0 and 11 bar, the pressure difference between outlet and inlet can be increased up to 5 bar. The rotor is driven by a 1.5 kW electric motor with speeds up to 50 rps. Several glass windows are mounted in the casing to enable visual observations and laser based flow measurements. The shaft is coupled to the motor on one side and the opposite side is manufactured as a hollow shaft which enables the electrical connection to a sensor that is placed inside the impeller. The impeller consists of eight blades, has a diameter of 130 mm and a clearance of 11 mm. The liquid flow enters the impeller in the axial direction and leaves it in the radial direction. The maximum gas flow rate is about 10 l/s, the maximum liquid flow rate depends on the backpressure and is 0.5 l/s for the present investigations. The separator shown in Fig. 2 acts also as a solid–liquid separator. The solid–liquid separation is carried out in a centrifuge mounted below the impeller. Solid particles are directed to the outer shell of the centrifuge by means of centrifugal forces and leave the separator continuously within a concentrated slurry flow. For the current investigations, only a two phase gas–liquid flow is considered. Investigations on three phase separation are reported on by Creutz (1998). Two outlets for the liquid are used as shown in Fig. 2. Outlet 1 is connected to the outer circumference of the impeller and resembles the typical configuration of a two-phase gas/liquid separator. Outlet 2 is the connection for the clean liquid leaving the centrifuge.

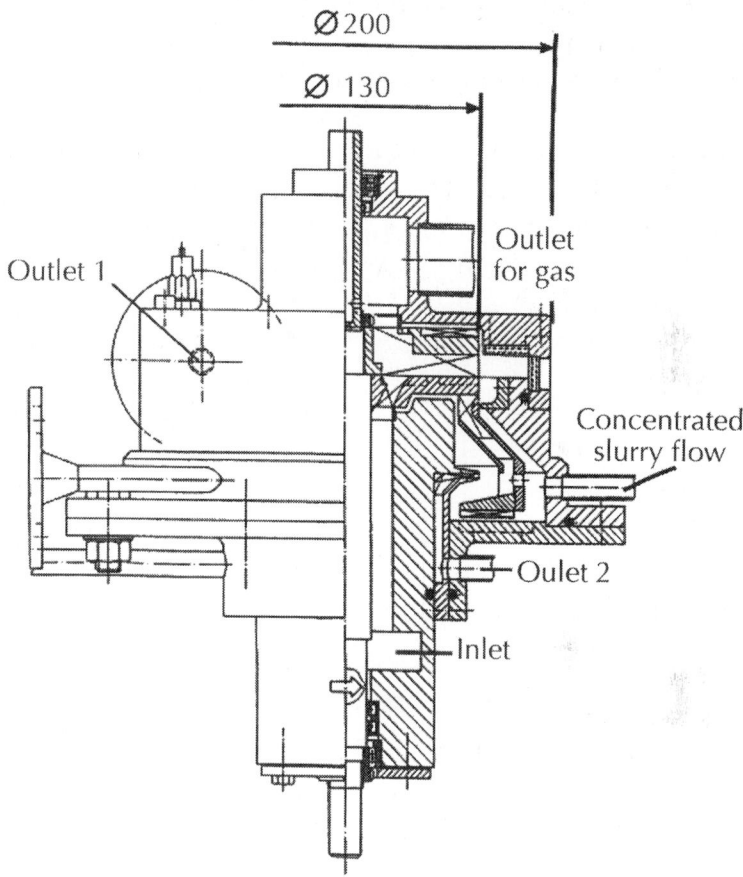

Figure 2: Technical drawing of the separator.

The experimental set up is shown in Fig. 3. The water-air two phase flow is fed into the separator via a 4 m long horizontal pipe (ID 16 mm). A transient slug flow is established in this pipe. The gas is separated and released to the atmosphere, the liquid leaves the separator either through outlet 1 or 2. The separated liquid may contain small amount of small gas bubbles. The mass flow and the mean density of the mixture is measured by a Coriolis flowmeter. Assuming no slip, this allows for the determination of the individual mass flow rates of gas and liquid. The liquid flow is sent back into the storage vessel where remnants of gas in the flow are removed by gravity. The total pressure in the system is controlled by a valve connected to the gas outlet. The pressure is also controlled inside the storage tank. The liquid flow rate is controlled by the pressure difference between the storage tank and the separator that corresponds to the two-phase pressure drop in the supply pipe while the gas flow rate is adjusted by a hand valve. The back pressure to the separator is regulated by valves in each of the liquid lines. The system can be operated with three-phase gas/liquid/solid flow as well.

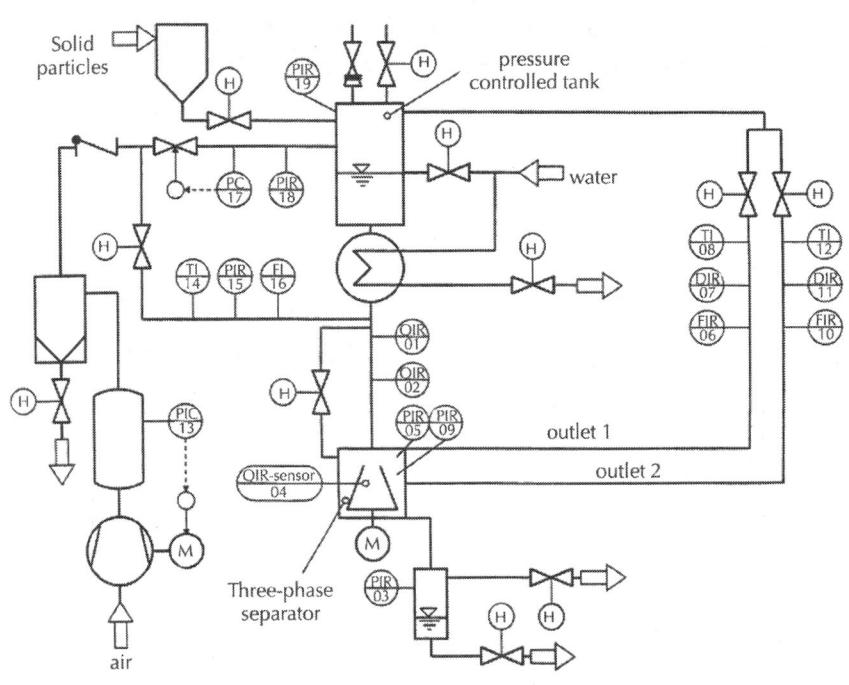

Figure 3: Schematic of the experimental set up.

MEASUREMENT TECHNIQUE

The system used for the measurement of the time-dependent variables is shown in Fig. 4. The flow at the inlet of the separator is a slug flow. The mean volumetric liquid flow rate is measured by the

Coriolis flowmeter at the outlet of the separator. The mean liquid fraction inside the cross section of the supply pipe is measured at two axially displaced positions via two pairs of ring electrodes with a frequency of up to 1 kHz. From these two measurements, the slug velocity of the two phase gas–liquid flow is calculated via a cross correlation between the two electrodes. From the liquid fraction, the slug velocity and the mean liquid flow rate, the dynamic liquid flow rate at the inlet of the separator is obtained (for details see Creutz, 1998). The back-pressure is measured by means of a pressure transducer. The liquid fraction inside the impeller is measured by means of an electrical resistance sensor as a function of the radial and the circumferential angle.

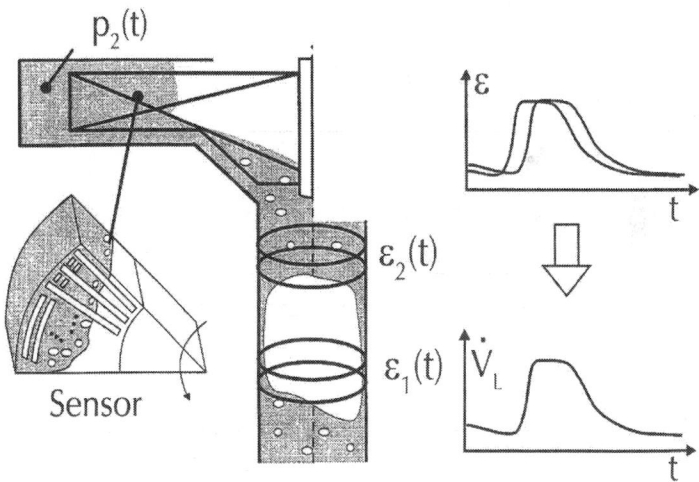

Figure 4: Schematic of the positions for the time-dependent measurements.

A schematic of the sensor and the electrodes is shown in Fig. 5. It is an integral part of the impeller and is located between two neighboring blades. Ten electrodes along the circumference and six electrodes along the radius of the impeller are used to measure an integral liquid fraction. The electrical potential of the counterpart of the electrode is grounded as well as all steel parts of the separator. The liquid fraction is a function of the electrical resistance between the electrode and the conducting counterpart of the sensor. The electrodes are surrounded by an electrical shield that provides a homogeneous electric field close to the electrodes leading to a linear relationship between liquid fraction and electrical resistance. Moreover, the electric shield damps the two-dimensional effects. The electrical potential in a radial cross section through the impeller resulting from a two dimensional finite-difference calculation is shown in Fig. 6. The highest current densities are obtained between the shield and the conducting material of the impeller. Here, the current density is approximately 10 times larger than inside the measurement cross section region, where the electrical field is nearly homogeneous.

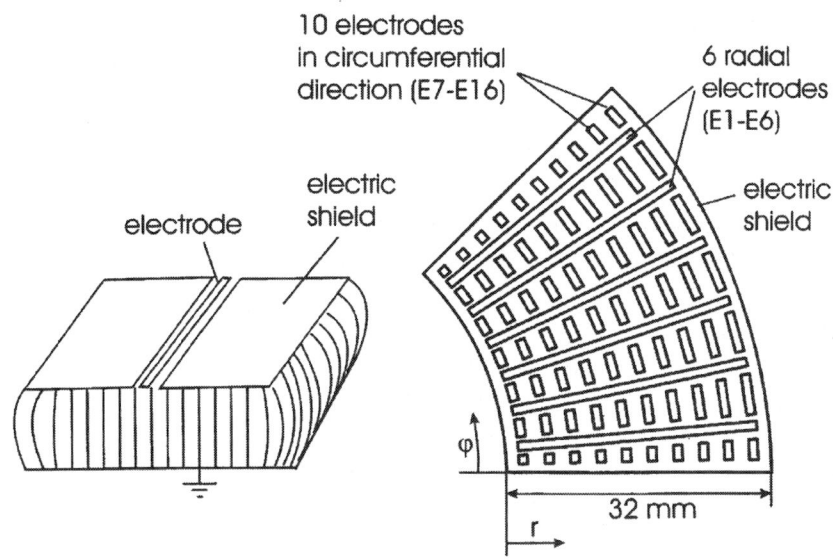

Figure 5: Schematic of the sensor and the electrodes.

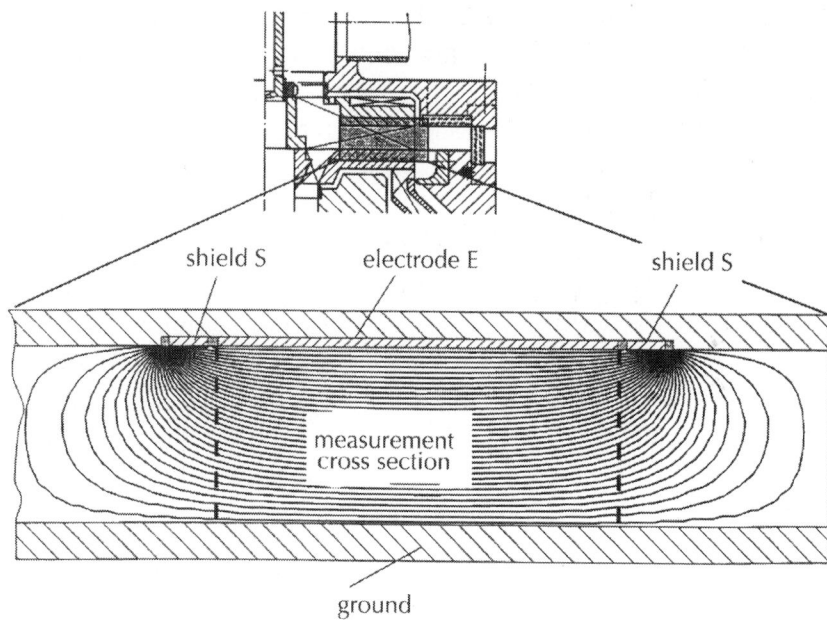

Figure 6: Iso-potential lines for a radial cross section through the sensor.

The electronic data acquisition of the sensor is shown schematically in Fig. 7. The electrodes are multiplexed by Siliconix SMD-analog multiplexers with an inner resistance of about 40 Ω. A 4-Bit binary code (A0-A3) to control the multiplexers is provided by a PC. The selected electrode (E) and the counterpart of the sensor (M) that is set to the ground potential are connected to a 250 Ω Wheatstone bridge. The bridge is supplied by 1.5 V AC voltage with a frequency of 5kHz. The signal of the bridge is amplified by a frequency shift amplifier, converted into a 12 bit digital signal and stored in the PC. The measurement frequency for a single electrode is 1 kHz and for a complete set of 16 electrodes it is 62.5 Hz. The electric shield is supplied with the alternating electrode potential, decoupled from the bridge by an operational amplifier. The multiplexers and the operational amplifier are embedded in the sensor and thus rotating with the impeller. The sensor is connected to a rotating measurement casing by cables through a drill-hole inside the shaft. High speed slip rings with an outer diameter of

0.8 mm are used for the electrical connection to the stationary parts of the data acquisition system. The metal structures of the separator are connected to the ground potential. The sensor is used for two different purposes. By using only one electrode in the radial direction, the position of the gas–liquid interface, $h*$ is measured with up to 1 kHz assuming a sharp interface. In combination with the back-pressure of the separator and the liquid flow rate, insight into the dynamic behavior of the system is obtained. These measurements are compared to a non-linear calculation in the following chapter. By using all electrodes, integral information on the liquid fraction inside the separator is obtained.

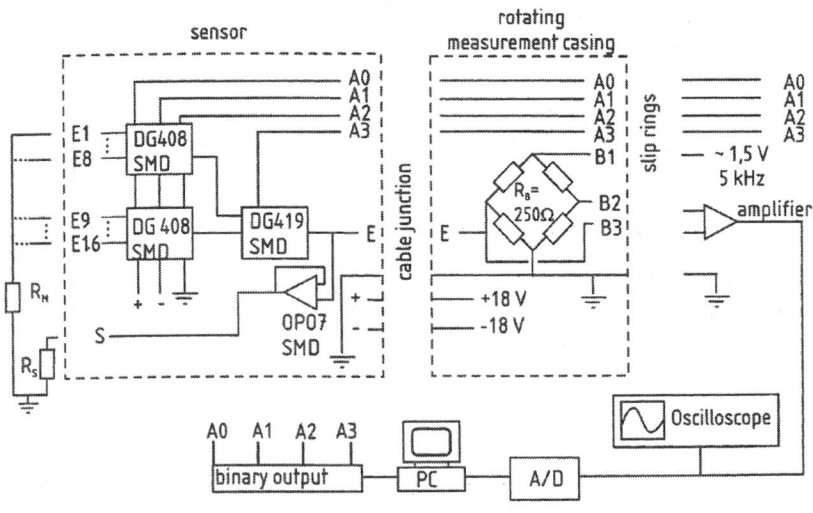

Figure 7: Data acquisition system of the sensor.

DYNAMIC BEHAVIOR OF THE SEP-ARATOR AND SCALE UP

The dynamic behavior of the gas–liquid separation inside the impeller is quantified with a theoretical approach. The formulation enables the calculation of three unknown variables, i.e. the back pressure of the separator, the liquid flow rate that leaves the separator and the position of the gas–liquid interface as functions of the liquid flow rate that enters the separator. A schematic of the gas–liquid separation and the relevant geometric quantities are shown in Fig. 8. The subscripts 1 and 2 indicate the separator inlet and outlet, respectively and the subscript 0 stands for ambient conditions. The following assumptions are made in the analysis:

- The friction inside the impeller is negligible.
- The position of the interface is one-dimensional.
- The liquid is completely removed from the gas and vice versa.
- The radial velocity of the liquid inside the impeller is approximately two orders lower than the tangential velocity and thus is neglected in the momentum equation.
- The angular velocity of the impeller is constant.
- Pressure drop due to acceleration is considered only in the outlet pipe.
- All balance equations are formulated in one dimension.

The three equations that are required for the mathematical solution are the mass and the momentum balance of the flow inside the impeller, and that for determining the pressure drop in the liquid outline. The mass balance of the flow through the impeller leads

$$\dot{V}_{L1} - \dot{V}_{L2} = \frac{dV_{L,FR}}{dt} = (R - h)2\pi x_{FR}\frac{dh}{dt}. \tag{1}$$

In this equation, V_{L1} is the volumetric liquid flow rate entering the separator, V_{L2} is the volumetric liquid flow rate that leaves the separator, R is the outer radius of the impeller, $V_{L,FR}$ is the volume of liquid inside the impeller, h is the position of the interface, x_{FR} is the clearance of the impeller and t is the time. Following the above assumptions, the integral form of the momentum equation inside the impeller in the radial direction yields

$$p_2 - p_1 = \rho_L\int_{R-h}^{R}\frac{v_U^2}{r}dr = \rho_L\int_{R-h}^{R}\omega^2 r\,dr = \frac{\rho_L}{2}\omega^2(2Rh - h^2) \tag{2}$$

with the pressure in the supply pipe, p_1 , the pressure behind the impeller, p_2, the density of the liquid, ρ_L, and the tangential velocity, v_U as a product of the radius r and the angular velocity ω. By the momentum equation, the position of the interface is dynamically correlated to the back pressure of the separator. The position of the interface adapts itself to a given pressure difference, (p_2-p_1) and, thus, inherently controls the system. For instance, by increasing the liquid flow rate entering the separator, the liquid level is increased. This leads to an increase of the outlet pressure and, thus, to an increase in the flow rate inside the following pipe. The liquid level adapts itself rapidly to variations in the liquid flow rate and allows for separating intermittent flows such as slug and plug flows. Friction, acceleration of the liquid and hydrostatic pressure drop are taken into account for calculating the pressure difference along the pipe:

$$p_2 - p_0 = \frac{\rho_L}{2} \frac{\dot{V}_{L2}^2}{R^4} \zeta_R + \rho_L \sum_i \frac{L_{R,i}}{A_{R,i}} \frac{\partial \dot{V}_{L2}^2}{\partial t} + \rho_L f \Delta h.$$
(3)

In this equation, p_0 is the ambient pressure, the pipe is characterized by the sum of i sectional relationships of length, $L_{R,i}$ and cross sectional area, $A_{R,i}$ as well as by its static height h. The friction factor of the pipe is defined as

$$\zeta_R = \frac{\Delta p_{Reib}}{\rho_{L/2} \dot{V}_{L2}^2} R^4.$$
(4)

The friction factor is obtained from experiments as a function of the liquid flow rate. Changing the friction factor by throttling the pipe, changes the mean position of the interface. High amounts of the acceleration force result in high peaks of the backpressure of the separator and shrinks the possible range of operation. The system of equations (, and) can be written in dimensionless form

$$\dot{V}_{L1}^* - \dot{V}_{L2}^* = (1 - h^*) \frac{\partial h^*}{\partial Fo},$$
(5)

$$Eu_2 - Eu_1 = \frac{1}{R^2} \int_{R-h}^{R} r \, dr = 2h^* - h^{*2},$$
(6)

and

$$Eu_2 - Eu_0 = \zeta_R \dot{V}_{L2}^{*2} + D_R^* \frac{\partial \dot{V}_{L2}^*}{\partial Fo} + H^*,$$
(7)

with the dimensionless position of the interface

$$h^* = \frac{h}{R},$$
(8)

the Fourier-number

$$Fo = \frac{t \omega R}{2\pi x_{FR}},$$
(9)

the dimensionless volumetric flow rate

$$\dot{V}^* = \frac{\dot{V}}{\omega R^3},$$
(10)

the dimensionless hydrostatic pressure

$$H^* = \frac{2g \Delta h}{\omega^2 R^2},$$
(11)

the Euler-number

$$Eu = \frac{2p}{\rho_L \omega^2 R^2}$$
(12)

and the differential coefficient

$$D_R^* = \frac{R^2}{\pi x_{FR}} \sum_i \frac{L_{R.i}}{A_{R.i}}.$$
(13)

The above system of equations is non linear and is solved numerically by a modified Runge–Kutta method. The results of these calculations will be compared to measurements in the following section. For the steady state layout of the separator and for determining the operating conditions, the position of the interface is calculated as a function of volumetric flow rate, the backpressure and the angular velocity of the separator. For that purpose, neglecting all derivatives to Fo yields the steady state solution of the system of equations

$$h_{\text{stat}}^{*} = 1 - \sqrt{1 - (\xi_{\text{R}} \dot{V}_{\text{stat}}^{*2} + H^{*})} = 1 - \sqrt{1 - Eu_{\text{stat}}} \tag{14}$$

with the steady state Euler-number

$$Eu_{\text{stat}} \equiv Eu_2 - Eu_1 = Eu_2 - Eu_0. \tag{15}$$

from

$$Eu_{\text{stat}} = \xi_{\text{R}} \dot{V}_{\text{stat}}^{*2} + H^{*}. \tag{16}$$

By linearizing , and around a steady state solution, a stability analysis is carried out. The linearization leads to

$$\dot{V}_{\text{L1}}^{+} = \dot{V}_{\text{L2}}^{+} + T_1 \frac{\partial \dot{V}_{\text{L2}}^{+}}{\partial Fo} + T_2^2 \frac{\partial^2 \dot{V}_{\text{L2}}^{+}}{\partial Fo^2}$$

$$K_1 \left(\dot{V}_{\text{L1}}^{+} + T_{\text{D}} \frac{\partial \dot{V}_{\text{L1}}^{+}}{\partial t} \right) = Eu_2^{+} + T_1 \frac{\partial Eu_2^{+}}{\partial Fo} + T_2^2 \frac{\partial^2 Eu_2^{+}}{\partial Fo^2} \}$$

$$K_2 \left(\dot{V}_{\text{L1}}^{+} + T_{\text{D}} \frac{\partial \dot{V}_{\text{L1}}^{+}}{\partial t} \right) = h^{+} + T_1 \frac{\partial h^{+}}{\partial Fo} + T_2^2 \frac{\partial^2 h^{+}}{\partial Fo^2} \tag{17}$$

with the linearized quantities that are indicated by '+' and defined in Section A.1. The system output (V_{L2}, p_2 and h) can be calculated independent of each other for a given excitation of the system (V_{L1}). The pressure at the entrance of the separator, p_1 and the ambient pressure, p_0 are set equal as they are in the experimental investigations. The differential Eq. (17) are of second order with the time constants T_1 and T_2. The differential part in the equations for the Euler-number and the position of the interface are indicated by the time constant T_D. The linearized system is stable. The damping coefficient of the linearized system is

$$D_{\mathrm{dyn}} \equiv \frac{T_1}{2T_2} = \frac{Eu_{\mathrm{stat}} - H^*}{\dot{V}^*_{\mathrm{stat}}\sqrt{2D_R^*}} = \frac{p_{2\mathrm{stat}}}{\rho_L \omega \dot{V}_{\mathrm{Lstat}}} \sqrt{\frac{\pi x_{\mathrm{FR}}}{2\sum_i L_{R,i}/A_{R,i}}}. \quad (18)$$

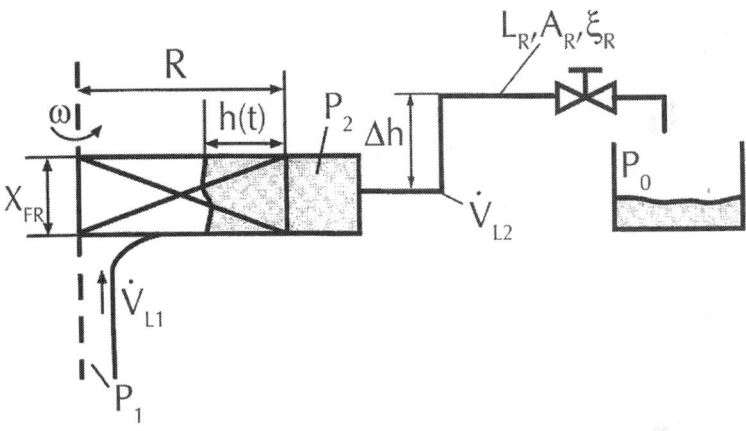

Figure 8: Schematic of the impeller and the relevant geometric quantities.

High values of the damping coefficient lead to smoothing large variations of the volumetric liquid flow rate. Technically, the damping coefficient is enlarged by small angular velocities, by wide impeller clearances, by a high ratio of backpressure to volumetric flow rate and by short pipes connected to the impeller. A long pipe additionally leads to increasing the differential coefficient T_D and thus to increasing pressure peaks following each slug that enters the separator.

EXPERIMENTAL RESULTS

The plot of the reduced steady state Euler-number versus the dimensionless liquid flow rate is shown inFig. 9. The theoretical prediction according to Eq. (16)is plotted as a line. The measurements are indicated as circles. They are based on variations of the separator's angular velocity, the volumetric flow rates of gas and liquid and on different kinds of liquid outlet pipes. The measurements resemble the theoretical prediction according to Eq. (12) with relative error of 5%. The reduced steady state Euler-number increases with the volumetric liquid flow rate to the power of two due to the turbulent flow in the liquid outlet pipe. This behavior elucidates the difference between the separator and a conventional centrifugal pump, where the back pressure usually decreases with increasing flow rate.

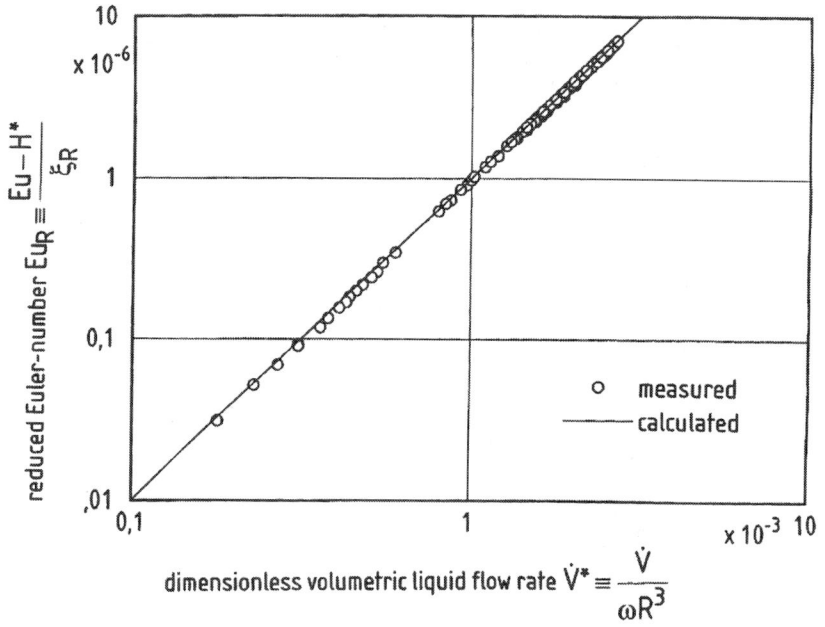

Figure 9: Reduced steady state Euler-number as a function of the dimensionless liquid flow rate.

In Fig. 10, the position of the interface obtained from experimental investigations is shown as a function of the Euler-number and is compared with theoretical findings for steady state flow (Eq. (14)). The position of the interface moves to the inside of the impeller with increasing Euler-number. Due to the sensor geometry, the position can be measured between $0.03 < h* < 0.54$. This region correlates with the Euler-numbers $0.06 < Eu < 0.79$. The measurements represented in the figure are obtained for liquid flow rates from 0.01 to 0.16 l/s and rotation frequencies between 8.5 and 50 Hz. The relative error of the measurements compared with Eq. (14) is ±20%. This high value is due to the assumption of an ideal interface. In reality the interface consists of a foamy zone between gas and liquid bulk phase as will be shown later.

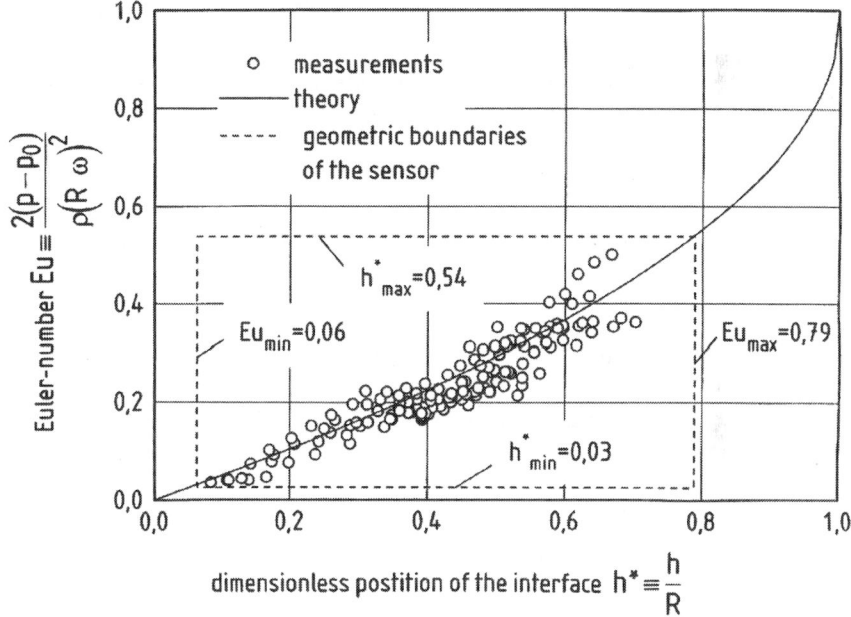

Figure 10. Dimensionless position of the interface as a function of the Euler-number.

In the upper diagrams of Fig. 11 the pressure at the outlet of the separator is plotted as a function of time for an intermittent flow at the inlet. The liquid flow rate at the inlet and at the outlet of the separator is plotted in the lower diagrams. The inlet flow rate is measured and the outlet flow rate is obtained by the calculations. The inlet pipe has an inner diameter of 16 mm and is mounted horizontally. Slug and plug flow exists with high time variations of the liquid flow rate. The mean flow rates of gas and liquid at the inlet to the pipe are kept constant. The damping coefficient is varied by changing the separator frequency. For a damping coefficient of 0.64 (left diagrams) the pressure reacts strongly on variations in the flow rate. By increasing the damping coefficient to 1.04 (right diagrams) the back pressure of the separator as well as the flow rate in the outlet are more even. The fluctuations in the outlet flow rate are much smaller that the fluctuations in the back pressure due to the missing differential part in the corresponding equation (Eq. (17)). They also decrease with increasing damping coefficient. This fact could also be observed visually at the reflux to the storage tank. Technically, the damping coefficient should be maximized when separating intermittent flows to minimize pressure fluctuations and transient liquid outflows.

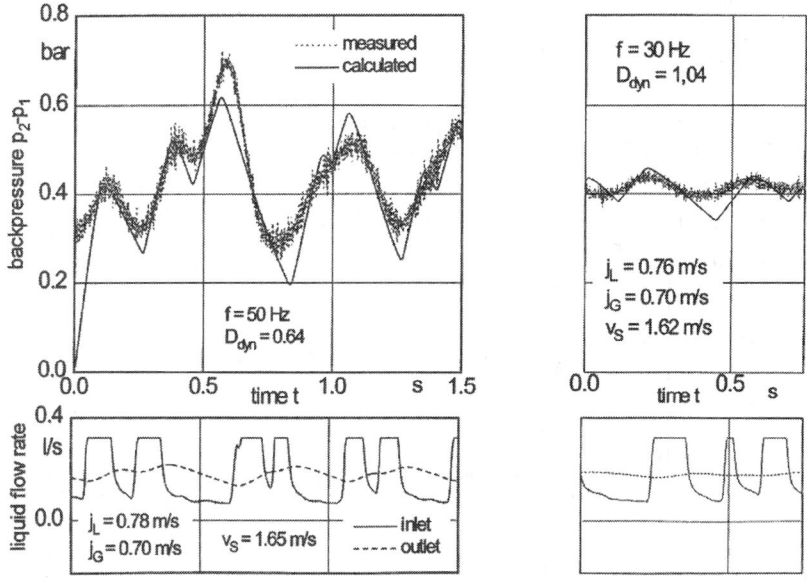

Figure 11: Pressure and the volumetric liquid flow rate as functions of the time for different damping coefficients.

By solving the system of non-linear differential equations (, and) with the measured liquid flow rate at the inlet as a boundary condition, the back pressure p_2 and the outlet liquid flow rate are calculated. The calculated values of the pressure match the experimentally obtained ones. Since back pressure and outlet liquid flow rate are coupled by a well known physically based equation (Eq. (3)), the behaviour of the outlet liquid flow seems to be reasonable, although the values are not experimentally verified. The physical basics deducted from the experiments are accounted for in the theoretical approach.

GAS–LIQUID SEPARATION

Gas is pulled into the liquid by the radial impulse of the liquid impinging into the interface. Some little amount of the gas penetrates through the liquid and leaves the impeller together with the liquid flow. Experimental investigations have been carried out on the gas–liquid separation inside the impeller. The volumetric gas fraction inside the ring chamber at the outer circumference of the impeller has been measured by removing the liquid directly behind the impeller (outlet 1, see Fig. 2). The volumetric gas fraction of the flow inside the centrifuge has been measured by removing the liquid through outlet 2. The gas fraction of the liquid flow is shown as a function of the position of the interface in Fig. 12 using the two different outlets. The rotational velocity of the separator, the pressure drop in the pipeline behind the separator and the liquid flow rate are varied. The gas fraction decreases with increasing position of the interface because gas bubbles have to penetrate through a wider liquid ring inside the impeller. The gas fraction is decreased significantly between the ring chamber and the centrifuge. At the inlet to the centrifuge, the flow is set to a solid body rotation. Fluctuations are damped and gas bubbles are separated and are removed together with a small amount of liquid through a gap between impeller and solid casing and penetrate back into the flow that enters the separator. If the position of the interface exceeds the radial position of the gas outlet ($h*=0.58$), liquid is carried through the gas outlet. Under common operation conditions, no liquid-carryover could be observed visually. In a technical application the position of the interface is varied by changing the rotational velocity of the impeller or by throttling the liquid pipe. If the entering flow is steady, the liquid level should be close to the maximum value. In case of an intermittent inlet flow, the level should be set to a value below the maximum to enable variations of the interface without liquid-carryover. Generally, no centrifuge is needed for gas–liquid separation. For this reason the following analysis concentrates on the flow inside the impeller only.

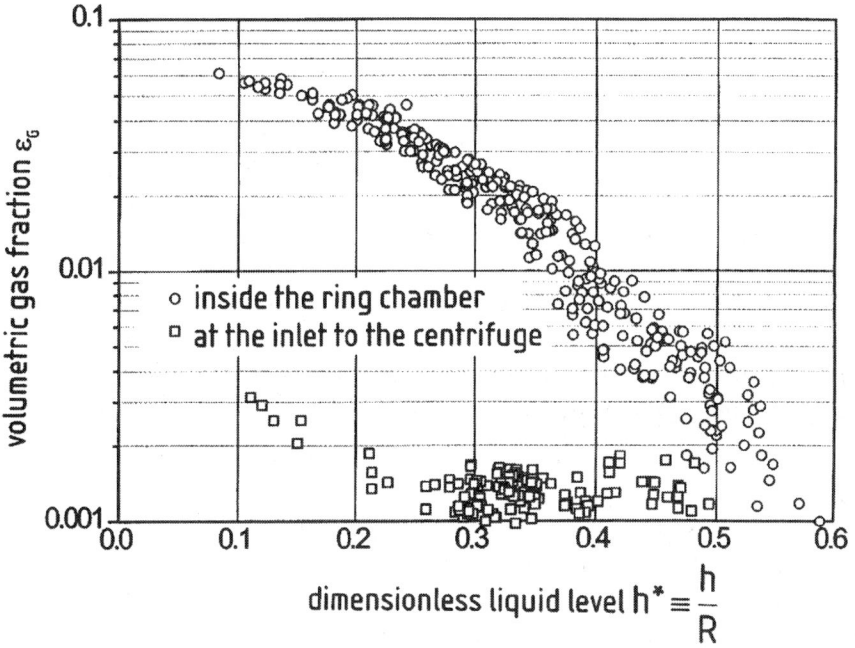

Figure 12: Volumetric gas flow rate in the liquid outlet as a function of the position of the interface.

In Fig. 13, plot of the liquid fraction and the velocity field in an axial cross section through the impeller is shown as a result of a numerical 3-D Euler–Euler calculation of the two-phase flow. The numerical calculations are based on a four field model, i.e. gas and liquid are assumed to be present in both dispersed fields (bubbles and droplets) and in the continuous fields. For the present calculations, the centrifugal forces, the buoyancy and the impulse forces (including added mass) play dominant roles and overcome the drag effects. Thus, the dependence of the solution on the bubble and droplet sizes, that are kept constant, is only little. The method is described by Creutz (1998) in detail. It is visible that the interface is not sharp but a foamy region extending approximately 1 cm. The flow field is three-dimensional but the position of the interface varies little with the circumferential angle.

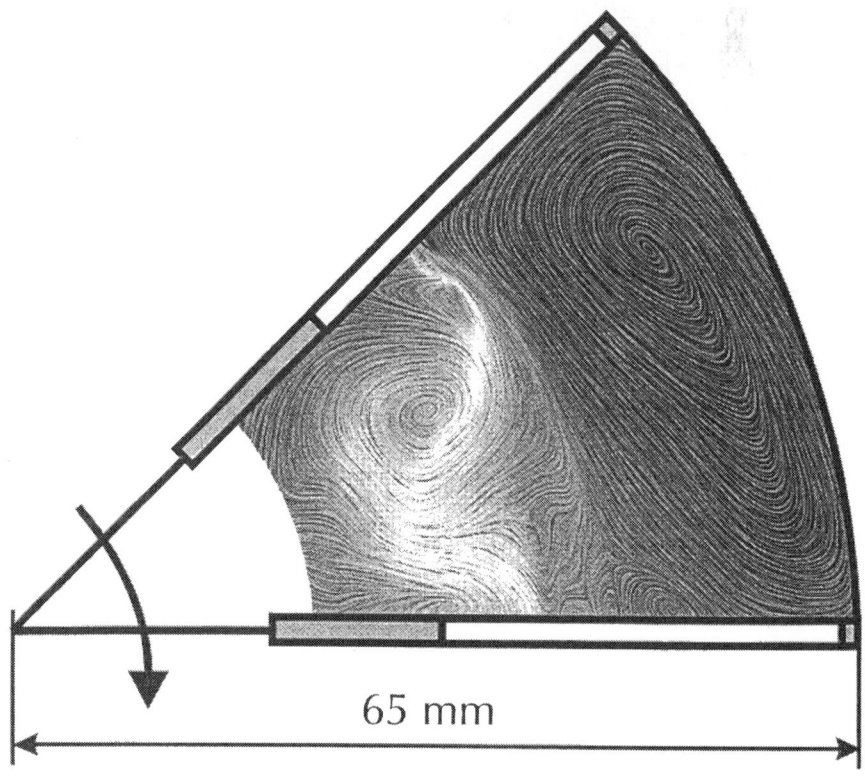

65 mm

Figure 13: Liquid fraction and the velocity field in an axial cross section through the impeller of the separator.

In Fig. 14 the mean liquid fraction that is integrated across the height and the circumference of the impeller is plotted as a function of the radius. The solution of the numerical calculation as well as experimental findings are shown. Experiments and calculations fit reasonable well except for a minimum liquid fraction in the continuous liquid region. This minimum characteristically occurs in all measurements and is explained by the trajectories of the bubbles through the liquid inside the impeller. The bubbles are initiated by the impingement of water on the gas–liquid interface with an initial velocity in the radial direction. The bubbles are decelerated, most bubbles stop and move back to the interface. The minimum liquid fractions indicates the position of low bubble velocity corresponding to the point of return of the majority of the bubbles.

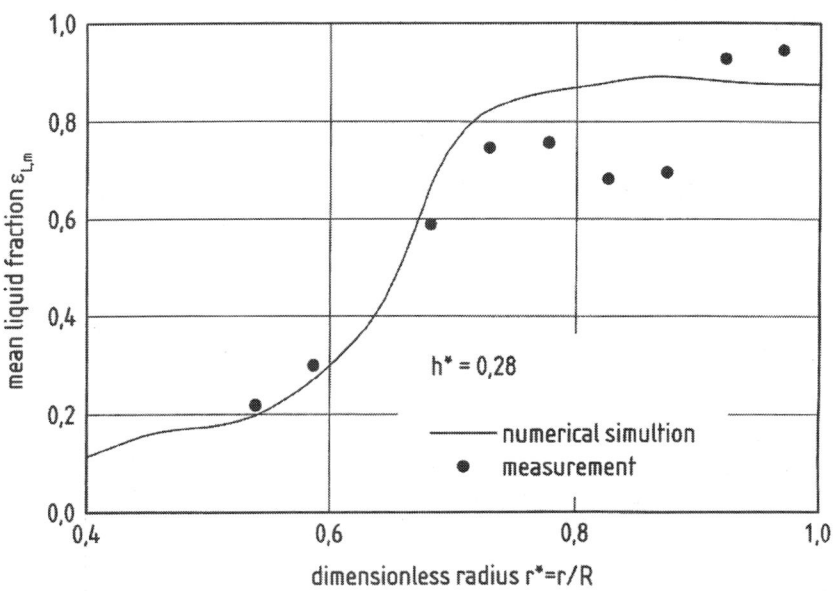

Figure 14: Mean liquid fraction across the height and the circumference of the impeller as a function of the radius.

A semi-empirical one-dimensional steady state model is proposed to predict the volumetric gas flow rate that leaves the separator through the liquid outlet. A plot of energy fluxes in the impeller is given in Fig. 15. The continuous gas and liquid regions are divided by the interface. Energy fluxes in the radial direction that are transported by the dispersed droplets and bubbles are shown schematically. The liquid droplets are accelerated due to centrifugal forces before reaching the interface. Their kinetic energy flux of the liquid in radial direction at the interface is:

$$\dot{E}_L = \frac{\rho_L}{2} \dot{V}_L \{v_{L,R}(h)\}^2.$$

(19)

In this equation, V_L is the volumetric liquid flow rate and $v_{L,R}(h)$ is the velocity of the liquid reaching the interface. The energy flux is transferred from the liquid to the gas-bubbles (including added mass) through the interface. The bubbles are decelerated by the centrifugal body force and the drag force. Depending on their terminal velocity, they pass the impeller (gas energy flux $\dot{E}_{G1}$) or they return to the interface and completely dissipate the gas energy flux $\dot{E}_{G2}$. The total terminal energy flux of the gas equals the energy flux of the liquid at the interface ($\dot{E}_L = \dot{E}_{G1} + \dot{E}_{G2}$). The ratio of the gas energy flux $\dot{E}_{G1}$ to the energy flux of the liquid is calculated by the energy flux ratio

$$X_E \equiv \frac{\dot{E}_{G,1}}{\dot{E}_L}.$$

(20)

The energy flux of the gas passing through the impeller is calculated by the kinetic energy of the gas bubbles leaving the interface:

$$\dot{E}_{G,1} = \frac{(\alpha \rho_L + \rho_G)}{2} \dot{V}_{G,1} \{v_{G,R}(h)\}^2$$

(21)

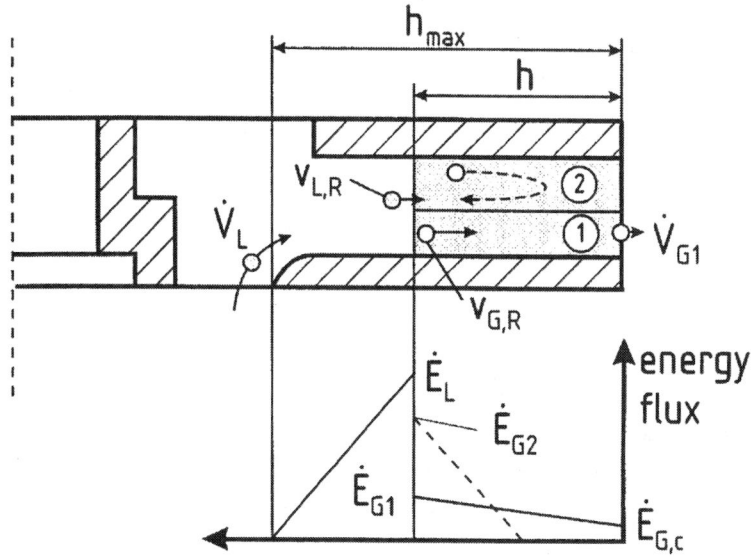

Figure 15: Schematic of the multiphase flow and the energy fluxes inside the impeller.

In this equation, $V_{G,1}$ is the volumetric liquid flow rate passing through the impeller, $v_{G,R}(h)$ is the radial terminal velocity of the corresponding gas bubbles at the interface and α is the added mass coefficient ($\alpha=0.5$). The volumetric gas flow rate $V_{G,1}$ is given by:

$$\dot{V}_{G,1} = X_E \dot{V}_L \frac{1}{\alpha + \frac{\rho_G}{\rho_L}} \left\{ \frac{v_{L,R}(h)}{v_{G,R}(h)} \right\}^2$$

(22)

or alternatively the gas fraction in the liquid outlet is given by:

$$\dot{\varepsilon} \equiv \frac{\dot{V}_{G,1}}{\dot{V}_{G,1} + \dot{V}_L} = \left\{ 1 + \frac{\alpha + \frac{\rho_G}{\rho_L}}{X_E} \left[\frac{v_{L,R}(h)}{v_{G,R}(h)} \right]^2 \right\}^{-1}$$

(23)

In these equations, the velocities of gas and liquid at the interface as well as the energy flux ratio are unknown. The latter one is assumed to depend on the layout of the impeller and its inlet only and will be obtained from experimental findings as an empirical constant. The velocities of the liquid at the interface is calculated assuming a constant centrifugal acceleration by integrating the momentum equation from the outer radius of the inlet to the position of the interface neglecting all frictional effects:

$$v_{L.R}(h) = (R\omega)\sqrt{(h^*_{max} - h^*)\left(1 - \frac{h^*_{max} + h^*}{2}\right)}.$$

(24)

In this equation, $h^*_{max} = 0.58$ is the position of the outer radius of the inlet. The derivation of the above equation is described in Section A.2. The terminal velocity of the gas bubbles necessary for passing through the impeller is obtained from ist momentum equation in the radial direction:

$$F_T = F_F + F_Z.$$

(25)

In this equation, F_T is the inertia force, F_F is the friction force and F_Z is the body force due to the centrifugal acceleration. The inertia force is

$$F_T = \frac{\pi}{6} d_B^3 (\alpha \rho_L + \rho_G) \frac{d^2 r}{dt^2},$$

(26)

the friction force is

$$F_F = -\frac{\pi}{4} d_B^2 \zeta_B \left(\frac{dr}{dt}\right) \left|\frac{dr}{dt}\right| \text{ with the friction factor } \zeta_B = \frac{24}{Re} = \frac{24 v_L}{\left|\frac{dr}{dt}\right| d_B}$$

(27)

for the Reynolds-number $Re < 1$. The centrifugal force is

$$F_T = \frac{\pi}{6} d_B^3 (\rho_G = \rho_L) r \omega^2.$$

(28)

In the above equations, d_B is the diameter of the bubble, r is the radius, t the time and v_L the kinematic viscosity of the liquid phase. Substituting , and in Eq. (25) leads to the differential equation to calculate the position of the bubble as a function of time:

$$\frac{d^2r}{dt^2} + C_A \frac{dr}{dt} + C_B r = 0$$

with

$$C_A \equiv 18 \frac{v_L}{d_B^2(\rho_G/\rho_L + \alpha)} \tag{29}$$

and

$$C_B \equiv \frac{\rho_L - \rho_G}{\alpha\rho_L + \rho_G} \omega^2.$$

That is solved to yield

$$r(t) = \frac{1}{e^{C_A t/2}} \left\{ \left(\frac{2v_0 + C_A r_0}{\omega_0} \right) \sin(\omega_0 t/2) + r_0 \cos(\omega_0 t/2) \right\} \tag{30}$$

with

$$\omega_0 \equiv \sqrt{4C_B - C_A^2} \quad \forall \quad 4C_B \geq C_A^2.$$

In this equation, v_0 is the terminal velocity of the gas bubble at the interface and $r_0 = R - h$ is the radial position of the interface. From this equation, the minimum terminal velocity $v_{0,min}$ is calculated that enables the bubble to pass through the impeller. For that purpose, the boundary conditions

$$\left. \frac{dr}{dr} \right|_{t=t^w} = 0 \quad \text{and} \quad r(t = t^w) = R \tag{31}$$

have to be satisfied. In these equations, t_w is the time the bubble needs to reach the outer circumference of the impeller. From the boundary conditions, the following equations are obtained:

$$2v_{0,\min} = \frac{C_A(2v_{0,\min} + C_A r_0) + \omega_0^2 r_0}{2v_{0,\min}\omega_0} \tan(\omega_0 t_w/2).$$

(32)

$$R = \frac{\left(\frac{2v_{0,\min}+C_A r_0}{\omega_0}\right)\sin(\omega_0 t_w/2) + r_0 \cos(\omega_0 t_w/2)}{e^{C_A t_w/2}}.$$

(33)

From these equations, the time t_w and the velocity $v_{0,\min}$ are calculated. In dimensionless form, the minimum terminal velocity is defined as

$$v_{0,\min}^* \equiv \frac{v_{0,\min}}{v_{0,\mathrm{gr}}} \equiv \frac{v_{0,\min}}{\omega\sqrt{2(R^2 - r_0^2)}}.$$

(34)

The dimensionless minimum terminal bubble velocity does not depend on the rotational velocity of the impeller. The bubble diameter is related to a characteristic bubble diameter

$$d_{B1}^* \equiv \frac{d_{B1}}{d_{B1,c}} \equiv \frac{d_{B1}}{3\sqrt{\frac{\nu_L}{\omega}}}\left\{\left(1 - \frac{\rho_G}{\rho_L}\right)\left(\frac{\rho_G}{\rho_L} + \alpha\right)\right\}^{1/4}.$$

(35)

The dimensionless minimum terminal bubble velocity is plotted vs the dimensionless bubble diameter in Fig. 16. With increasing diameter, the influence of the friction diminishes, the minimum velocity remains constant. Here, the analysis becomes independent of the bubble diameter. Smaller gas bubbles tend to return to the interface and are accounted for by the energy flux ratio. It is assumed that the diameter of all gas bubbles passing through the impeller is d_{B1}^* and that the terminal velocity of these bubbles equals the minimum terminal velocity $v_{0,\min}^* = 1$ leading to

$$v_{G,R}(h) = \omega\sqrt{2(R^2 - r_0^2)}.$$

(36)

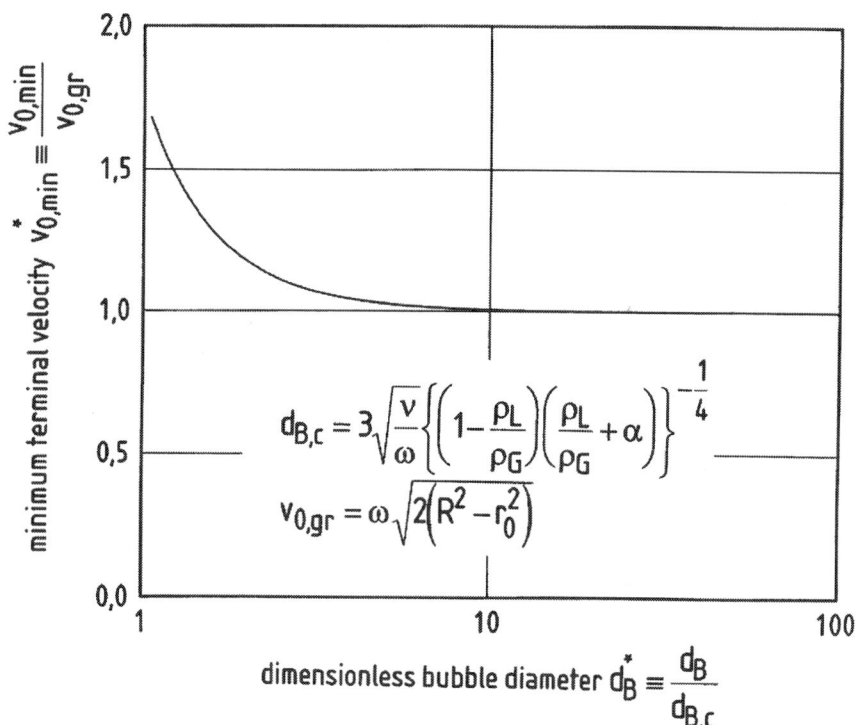

Figure 16: Minimum terminal bubble velocity as a function of the dimensionless terminal velocity of the gas bubbles.

Experimental findings of the gas energy flux $\dot{E}_{G,1}$ are plotted versus the liquid energy flux $\dot{E}_L$ in Fig. 17 . The volumetric flow rates of gas and liquid are varied as well as the back pressure and the rotational velocity of the impeller. The gradient of the plot corresponds to the energy flux ratio. For separation frequencies $f \geq 40$ Hz the error of the model is approximately ±20%. For small separation frequencies, the effect of friction on the trajectory of the gas bubble can not be neglected leading to a over-prediction of the energy flux $\dot{E}_{G,1}$. The energy flux ratio is assumed to characterize the geometric quality of the impeller and can technically be minimized by further optimizing the inlet section of the impeller.

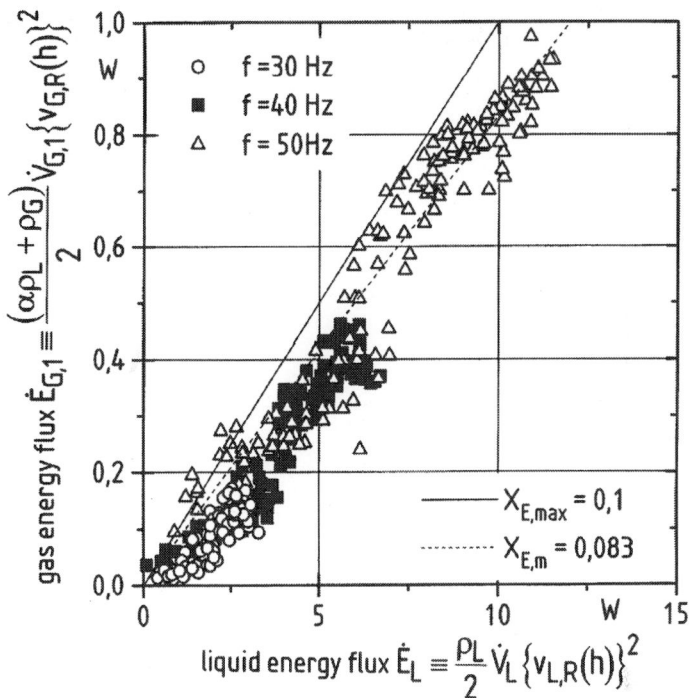

Figure 17: Energy flux $\dot{E}_{G,1}$ as a function of the energy flux $\dot{E}_L$.

CONCLUSIONS

A separator has been presented that enables the separation of gas and liquid for highly transient slug flows. The dynamic behavior of the separator has been modeled based on a simple physical approach without empirical factors. The theory is compared to experimental findings. The quality of gas–liquid separation is described using a semi-empirical model. This model leads to the determination of an empirical constant that stands for the quality of the impeller. Based on a linearized theory, the scale up of the separator is enabled.

ACKNOWLEDGMENTS

The authors gratefully acknowledge the German DFG, SFB 264 for the generous financial support.

APPENDIX A

A.1. Appendix A1

Linearized quantities and constants:

$$\dot{V}_{L1}^{+} = \dot{V}_{L1}^{*} - \dot{V}_{Lstat}^{*}, \quad \dot{V}_{L2}^{+} = \dot{V}_{L2}^{*} - \dot{V}_{Lstat}^{*}, \quad |Eu_{2}^{+} = Eu_{2}^{*} - Eu_{2stat}^{*}, \quad h^{+} = h^{*} - h_{stat}^{*},$$

$$T_{1} = \xi_{R}\dot{V}_{stat}^{*}, \quad T_{2} = \sqrt{\frac{D_{R}^{*}}{2}}, \quad T_{D} = \frac{D_{R}^{*}}{2\xi_{R}\dot{V}_{stat}^{*}} \quad K_{1} = 2\xi_{R}\dot{V}_{stat}^{*}, \quad K_{2} = \frac{\xi_{R}\dot{V}_{stat}^{*}}{(1 - h_{stat}^{*})}.$$

$$(A1)$$

A.1.1. Appendix A2

Derivation of (Eq. (24)): The velocity of the liquid phase that reaches the interface is obtained from the momentum equation of a liquid droplet. Frictionless motion of the liquid phase is assumed. The droplet is accelerated due to the centrifugal force corresponding to the rotational velocity assuming a solid body rotation inside the impeller:

$$\frac{dv}{dt} = \frac{d^2 r}{dt^2} \approx R_{m}\omega^2 \quad \forall \quad |r - R_{m}| \ll R_{m}.$$

$$(A2)$$

In this equation, v is the radial velocity of the droplet, r is the radius, R_{m} is a mean radius of the droplet motion and ω is the angular velocity. Integrating this equation two times leads to the radius of the droplet as a function of the time.

$$r = \frac{R_{\mathrm{m}}}{2}\omega^2 t^2 + R - h_{\max} \tag{A3}$$

For $t=0$ s the radius of the droplet equals the outer radius of the inlet into the impeller $r_0=(R-h_{\max})$. At the time $t=T$ the droplet impinges on the interface with the radius $r(t=T)=R-h$. The time T is obtained by

$$R - h = \frac{R_{\mathrm{m}}}{2}\omega^2 T^2 + R - h_{\max} \quad \Leftrightarrow \quad T = \sqrt{\frac{2(h_{\max} - h)}{R_{\mathrm{m}}\omega^2}} \tag{A4}$$

and the impinging velocity of the droplet by

$$v(t = T) = \sqrt{2R_{\mathrm{m}}\omega^2(h_{\max} - h)}. \tag{A5}$$

Assuming the mean arithmetic radius R_{m}

$$R_{\mathrm{m}} \approx R - \frac{h + h_{\max}}{2} \tag{A6}$$

And taking into account the definition of $h*$ (Eq. (8)), Eq. (41) corresponds to Eq. (24).

REFERENCES

1. Creutz, M., 1998. Separation dreiphasiger instationaÈr stroÈmender Gemische. VDI Fortschritt-Berichte, VDI Verlag-DuÈsseldorf, Reihe 3, nr. 521, ISBN 3-18-352103-2.

2. Hollenberg, J.F., de Wolf, S., Meiring, W.J., 1995. A method to suppress severe slugging in ¯ow line riser systems. 7th International Conference on Multiphase Production. 7±9. Cannes, France.

3. Kouba, G.E., Shoham, O., Shirazi, S., 1995. Design and performance of gas liquid cylindrical cyclone separators. 7th International Conference on Multiphase Production. 307±327. Cannes, France.

4.   Muschelknautz, S., Mayinger, F., 1990. StroÈmungsun tersuchungen in der Austrittsleitung und in einem Zyklon

5.   abscheider bei Druckentlastung. Chemie-Ingenieur-Technologie 62 (7), 576±577. Nebrensky, J.R., Morgan, G.E., Oswald, B.J., 1980. Cyclone for gas/oil separation. International Conference on Hydrocyclones. 167± 178. Cambridge, UK.

Chapter 8

Influence of the Baffle Clearance Design on Hydrodynamics of a Two Riser Rectangular Airlift Reactor with Inverse Internal Loop and Expand-ed Gas–Liquid Separator

Peter M. Kilonzo[a], Argyrios Margaritis[a],
M.A. Bergougnou[a], JunTang Yu[b], and Ye Qin[b]

[a]Department of Chemical & Biochemical Engineering, University of Western Ontario, London, Ont., Canada N6A 5B9

[b]Biochemical Engineering Research Institute and State Key Laboratory of Bioreactor Engineering, East China University of Science and Technology, 130 Meilong Road, Shanghai 200237, China

ABSTRACT

The influence of baffle clearance design on the liquid circulating velocity, gas holdup and pressure drop in a two riser rectangular airlift reactor with inverse internal loop and expanded gas–liquid separator was investigated using water and mineralized CMC solutions covering a range of effective viscosity from 0.02 to 0.5 Pa s and surface tension from 0.065 to 0.085 N/m. The gas holdup results in the riser, downcomer, and gas–liquid separator were satisfactorily derived using expressions obtained via dimensional analysis. The separator gas holdup was found to be similar to the total gas holdup in the airlift reactor. The baffle clearances were found to influence the liquid circulation velocity to some degree, with the bottom clearance being the significant design parameter. An attempt was also made to correlate the liquid velocity using empirical equations of the loss coefficient in the baffle top and bottom zones. The calculated and observed liquid circulation velocity agreed well with an error of ±29% for the air–water system.

INTRODUCTION

The number of attractive features of airlift reactors have led to increasing usage of these contactors in environmental remediation technology, the chemical process industry and the biotechnology-based manufacture [1] and [2]. Airlift reactors have an established niche in high-strength activated type treatment of wastewater where the high-oxygen transfer capability, low power requirements and non-mechanical agitation are particular advantages of these systems [3] and [4]. The gas holdup difference causes liquid circulation flow, which is a characteristic behavior in all airlift reactors. In most cases, gas is also circulated since small bubbles are easily entrained into a down comer by the liquid down flow.

Both gas holdup and liquid circulation velocity parameters govern the oxygen transfer from the gas phase to the liquid phase and the homogeneity of the airlift reactor, respectively [5].

The baffle bottom clearance (spacing between the lower end of the baffle and the base plate of the reactor) determines the rate of liquid and gas circulation through the loop. The distance between the upper end of the baffle and the liquid level (the top clearance) determines the amount of liquid/gas in the gas–liquid separation zone, which is above the riser (draft tube sparged) or down comer (annulus sparged).

The geometry of this region alone determines, to a large extent, the proportion of gas that is recycled through the down comer. Thus, the design of the baffle clearances of an airlift reactor affects the gas holdup difference between the riser and down comer; hence, the driving force for liquid circulation is affected. Only limited studies exist in the current literature of the effect of baffle clearance design on hydrodynamic performance characteristics of airlift reactors [6], [7], [8], [9], [10], [11], [12], [13], [14] and [15].

This work reports on the effects of baffle clearance design, on gas holdup and liquid circulation velocity in a two riser rectangular airlift reactor with inverse internal loop (annulus sparged) and expanded gas–liquid separator. Understanding these hydrodynamics is essential to successful design of airlift reactors for environmental engineering and other applications. Its particularity lies in the fact that it is a two riser airlift reactor.

The rectangular shape presents good mixing and better mass transfer performance [9], [16], [17],[18], [19], [20], [21] and [22]. Moreover, this shape was chosen because of its application in wastewater treatment: a third phase can be added to be used as microorganism support. Bigger rectangular plants are easier to build than their cylindrical counter parts [23] and [24].

EXPERIMENTAL

Reactor

The reactor apparatus for investigating the effect of baffle clearance design on the reactor hydrodynamic performance is shown in Fig. 1. The reactor was constructed of Plexiglas (poly-methyl methacrylate). It consisted of five sections: a gas–liquid separator, two risers, a downcomer and a bottom section. Each of the two risers and the downcomer shared an adjustable straight rectangular baffle (0.088 m × 1.056 m) as a central wall. The baffles were fitted to the column walls using a strip of rubber between them and the walls to ensure that no fluid traverses either from the riser into the downcomer and vice versa. The bottom clearance h_b between the baffle bottom and the base plate was varied from 0.01 to 0.1 m by adjusting the vertical position of the baffles on the column walls using stainless steel screws. The dimensions of the riser (s) and the downcomer given in Fig. 1 make the ratio of the cross-sectional area of the downcomer to the one of the risers equal to 1.33. The static gas-free height H_L of the liquid was varied in the range 1.0–1.4 m, giving baffle top clearance h_t in the range of 0–0.30 m for the range of gas velocity used in this study. The aspect ratio ($H_{r/d}/D_h$) of the reactor was 10. This ratio value was based on the riser and downcomer height (neglecting the liquid level in the separator) and the hydraulic diameter (D_h) of the riser and downcomer [25] and [26].

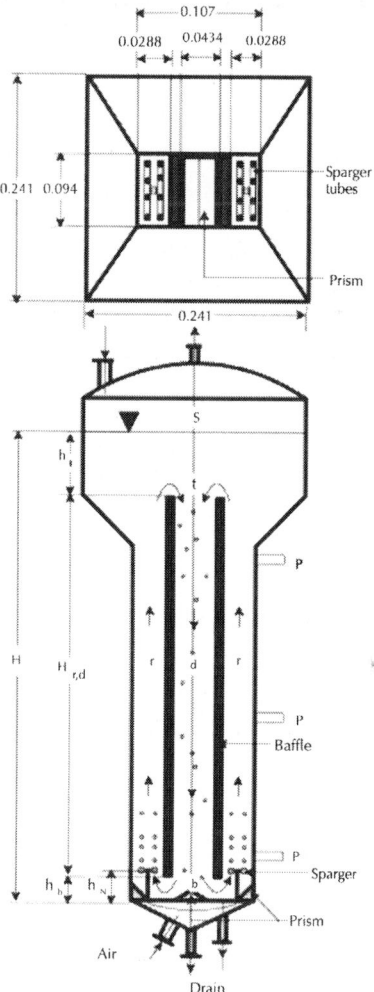

Figure 1: Dimensioned elevation sketch of rectangular-column airlift reactor used. Symbols: h_t, top clearance; h_b, bottom clearance, h_N sparger clearance; r, riser, d downcomer, s separator, p pressure tap, b bottom, H, height, $H_{r,d}$ riser/downcomer height, t top (all dimensions in m).

Two-pair ladder-type air spargers were used consisting of perforated plastic glass pipes of i.d. 0.04 m. The spargers were 0.018 m just above the bottom end of the baffles. A total of 138 sparger holes each of which having diameter D_o = 4.0 m × 10^{-4} m and equidistantly spaced were used for air distribution, giving a free area of 0.32% of the total cross-section area of the risers. This arrangement gave satisfactory gas distribution over all the orifices of the sparger at the lowest gas flow rates. For the latter, the requirement that the Weber number (W_{eo}):

$$W_{eo} = \frac{U_{og}^2 D_o \rho_{og}}{\sigma_L} > 2$$

(1)

based on the orifice diameter be greater than 2 was satisfied. This was as a result of the small range of flow rates used in this work. Nevertheless, due to the small diameter of the orifices, no weeping was observed even when some of the orifices were not active. The vertical position (h_N) of the spargers above the base plate was allowed to move as the baffle bottom clearance was varied.

Plastic prism inserts placed at the base plate (cf., Fig. 1) eliminated stagnant zones and ensured smooth movement of the fluid from the downcomer into the riser.

Systems

All experiments were carried out with air. Gas was sparged into the annulus between the column and the riser. Water and mineralized carboxy-methyl cellulose (CMC-Sigma Sodium salt, Medium Viscosity, No. C-4989) aqueous solutions were used as the liquid phase. The physical properties of the experimental media for which the data points were obtained are summarized in Table 1.

Table 1: Properties of experimental media used for data presentation

Medium	Density	Surface tension	Power-law parameters	
	ρ_L (kg/m³)	σ_L (N/m)	n (–)	K (Pa sⁿ)
De-ionized water	1000	0.0735	1.000	0.001
0.15 M NaCl–0.25% CMC solution	1004	0.0740	0.902	0.016

0.15 M NaCl–0.5% CMC solution	1007	0.0750	0.864	0.035

Air was sourced from a laboratory compressor via a pressure regulator, needle valve and rotameter that facilitated precise adjustment of gas flow rate. Compressed and oil-free air was used as the gas phase in all experiments. The air superficial velocity was varied from 0.03 to 0.25 m s^{-1}.

Measurements

The local gas holdups in the riser and the down comer were determined manometrically [1]. The riser and down comer each had three manometer/pressure taps located at 0.107 m (bottom port), 0.515 m (middle port) and 0.910 m (top port), respectively, from the base plate and were connected to inverted three-tube type water manometers. For each operating condition, a profile of E_g versus the height z in each region could be drawn according to the following expression [8]:

$$E_{gr,d} = 1 - \left(\frac{\Delta P_{r,d}}{\rho_L g \Delta z} \right) \tag{2}$$

The overall gas holdup (E_T) was determined by visual measurements of the clear liquid height (H_L) and the aerated liquid height (H_D) according to the following equation:

$$E_{gT} = 1 - \frac{H_L}{H_D} \tag{3}$$

The gas holdup in the gas–liquid separator was calculated from a gas volume balance:

$$EgTVgT = EgrVgr + EgdVgd + EgsVgs + EgbVgb \tag{4}$$

Neglecting the last term in the right-hand side in view of the small volume and very low gas holdup in the bottom section and taking into account the geometrical characteristics of the system, the separator holdup may be calculated from the equation [27]:

$$E_{gs} = \frac{(E_{gT}V_{LT}/(1 - E_{gT})) - E_{gr}H_r A_r - E_{gd}H_d A_d}{(V_{LT}/(1 - E_{gT})) - H_r(A_r - A_d) - h_b A_b}$$

(5)

Liquid Circulation Velocity

Liquid circulation velocity was determined by a signal-response technique using HCl acid tracer and a pH electrode detector [28], [29], [30] and [31]. In this work, four pH electrodes, two located in the riser and two in the downcomer sections were used. The response time of the pH electrodes was 0.1 s. The response of a pulse input of the tracer was signaled by the pH electrodes, and monitored by a pH meter (sensitivity: 0.01), respectively, and then simultaneously recorded every 5 s on a chart and a digital printer. The chart recording was synchronized manually with the introduction of every pulse of 10 mL 4 M (4N) HCL solution through the injection port. For the pH measurements in the riser section, the bottom port was used as the tracer injection port. The top port was used as tracer injection port during pH measurements in the downcomer section.

A typical response curve of pH against time obtained by the pH electrodes is shown in Fig. 2. The measured pH values of the liquid in the riser and the downcomer sections were then converted to the actual concentrations through a calibration curve.

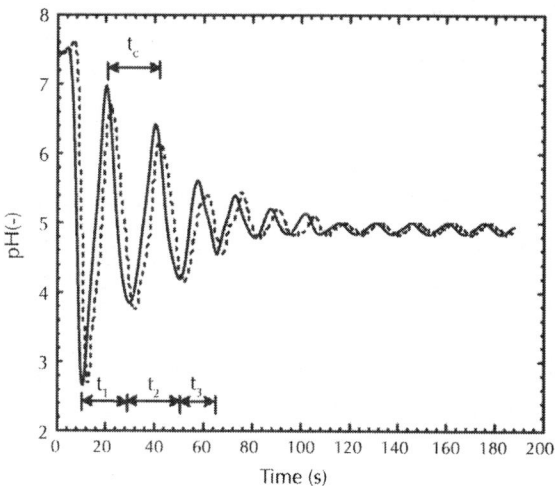

Figure 2: Typical response curves measured using two pH electrode sensors, one located at the entrance and another at the exit of the riser or downcomer regions of the airlift reactors.

The mean circulation time, t_C (time for a liquid volume element to travel once around the riser–downcomer circuit) was determined directly from the response curves observed for each airflow rate. The number of cycles detectable varied with the airflow rate. At low airflow rate, the number of cycles was six but at the highest airflow rate it was only a few. Therefore, experiments conducted at high-airflow rates were duplicated to obtain more reliable values for t_C.

With the mean circulation time t_C, the liquid linear and superficial velocities in the riser and downcomer were estimated using the following equations:

$$U_{Lr} = \frac{z_r(1 - E_{gr})}{\Delta t_r}$$

(6)

$$U_{Ld} = \frac{z_d(1 - E_{gd})}{\Delta t_d}$$

(7)

where z_r and z_d are the distances between the two pH electrodes in the riser and downcomer, respectively, and Δt_r and Δt_d are the average differences in response time of the second and third peaks of the response curves obtained by the two pH electrodes in the riser and downcomer, respectively. Note that the second and the third peaks of the response curves are used to obtain Δt_r and Δt_d because the first peak of the response curve obtained by the lower pH electrode in the riser is not well-established at higher aeration rate [5].

RESULTS AND DISCUSSION

Gas Holdup

Fig. 3 describes the effect of the bottom clearance on riser and downcomer gas holdups for the air–water system. The gas holdup increases in the riser as the bottom clearance decreases. This can be understood as being caused by a decrease in the liquid velocity. This is especially evident in the data for the downcomer gas holdup, since for the smallest bottom clearance the velocity of the liquid is so restricted by the pressure drop in the bottom of the reactor that almost all the gas disengages in the gas–liquid separator. The gas holdup in the downcomer is almost zero at low values of the superficial gas velocity U_{gr}.

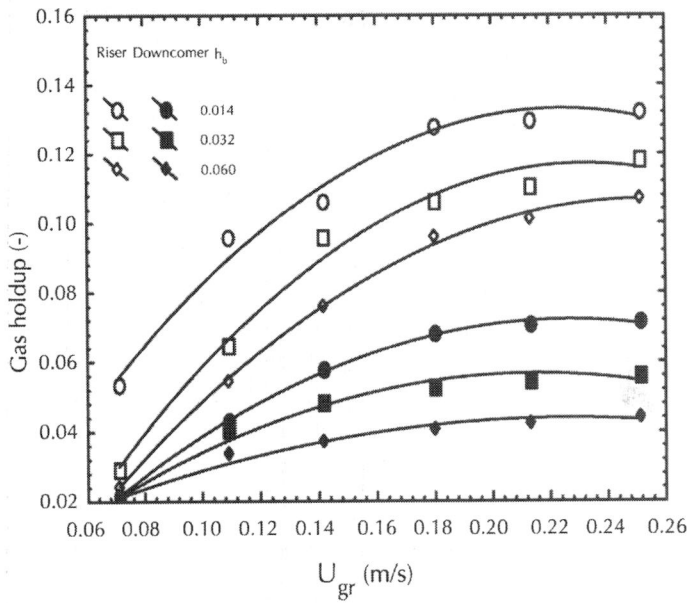

Figure 3: Average riser and downcomer gas holdup for the bottom and top clearance of $h_t = 0.225$ m; air–water system.

Fig. 4 shows the gas holdup in the riser and downcomer, for a bottom clearance of 0.014 m and two values of the top clearance: $h_t = 0.225$ and 0.265 m. Data are presented for water and 0.15 M NaCl–0.5% CMC solution. While for water h_t has a small effect, this is not so for the more viscous mineralized CMC solution. In the latter, the lower rising velocity of the bubbles causes more of them to be entrained and carried down by the liquid, and therefore the larger residence time in the gas–liquid separator region due to the larger bubbles that recirculate.

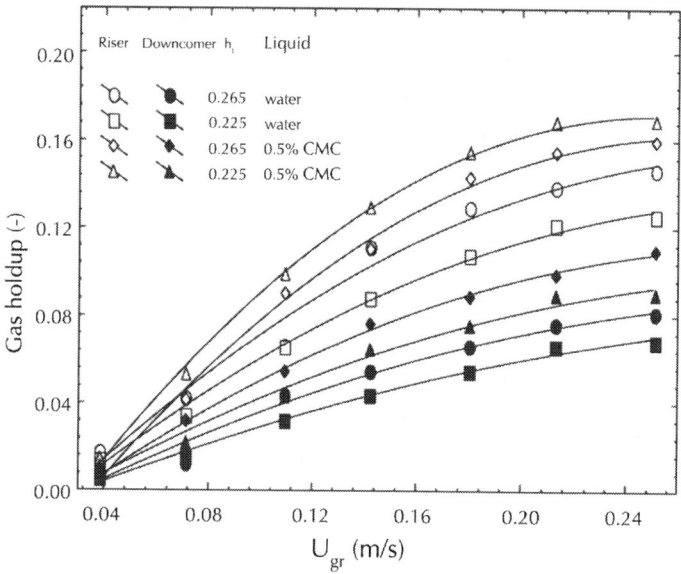

Figure 4: Average riser and downcomer gas holdups for a bottom clearance of $h_b = 0.014$ m and top clearances of $h_t = 0.225$ and 0.265 m.

The lower h_t gives a shorter residence time in the gas–liquid separator, a larger bubble recirculation, and, therefore, a larger gas holdup. This is especially evident in the downcomer for the mineralized viscous CMC solution.

The measured total and separator gas holdups are shown in Fig. 5 as a function of the superficial gas velocity U_{gr} for different liquids and constant values of the bottom and top clearances. Fig. 6 shows separator gas holdup as a function of the superficial gas velocity U_{gr} for various values of the bottom and top clearances. It is interesting to note that the separator gas holdup is essentially the same for $h_b = 0.032$ and 0.060 m (cf., Fig. 6), indicating balancing of the decrease of E_r by an increase of E_d as the liquid velocity increases. The separator gas holdup (E_s) was higher for the 0.25% and 0.5% NaCl–CMC solution than for water, especially at high U_{gr}. This is due to the lower bubble disengagement, which in turn results from the lower rising velocity in the mineralized viscous CMC solution. The separator gas holdup behaves generally as the total gas holdup

(E_T) (cf., Fig. 5). Comparing Fig. 3 and Fig. 4 one can see the effects of the expanded gas–liquid separator. All the data indicates that disengagement was more effective in the rectangular airlift reactor proposed in this study. This indicates that the cross-sectional area of the gas separator has a strong influence on the fluid dynamics of the reactor. This was shown by the regression analysis, which is presented in Section 3.2.

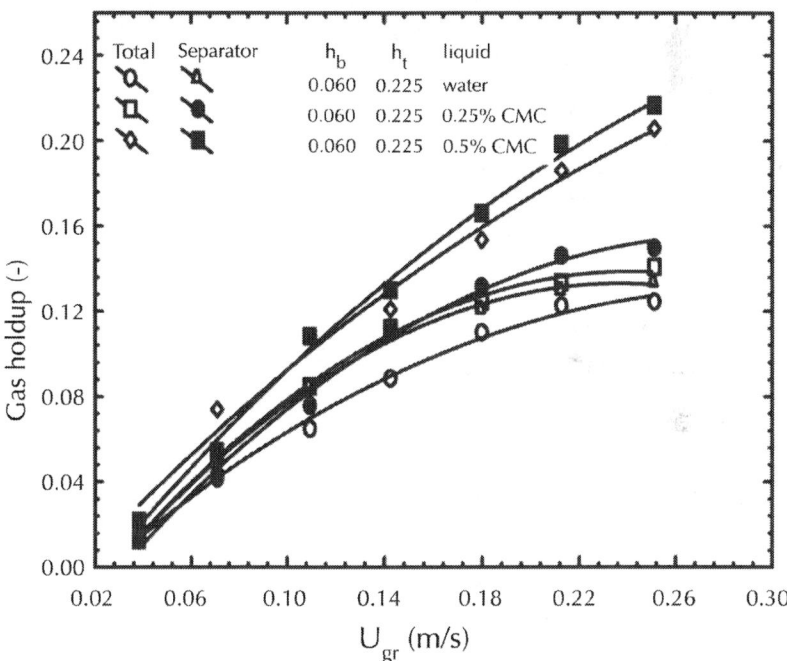

Figure 5: Total and separator gas holdup for constant top and bottom clearances and three liquids.

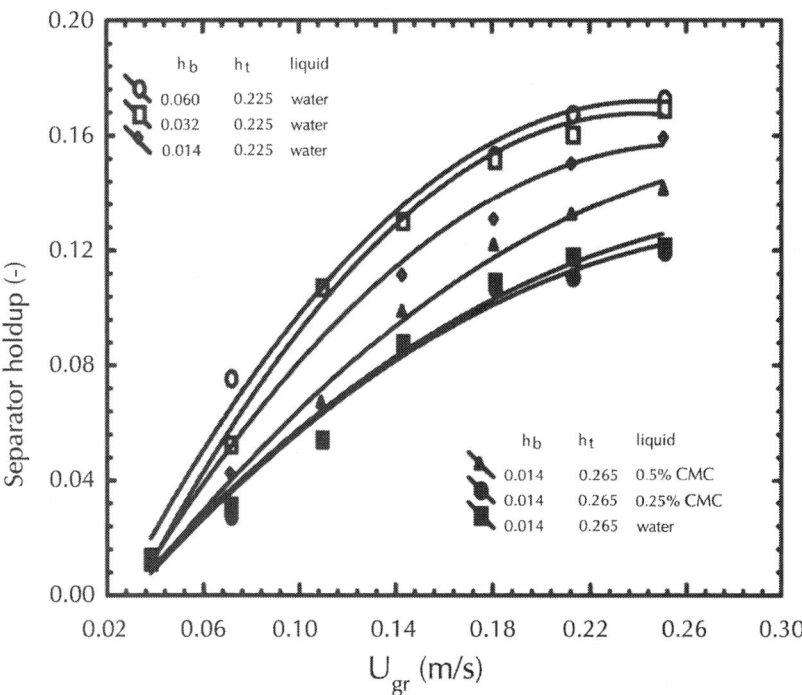

Figure 6: Separator gas holdup for various top and bottom clearances and three liquids.

Liquid Circulation Velocity

Fig. 7 shows the liquid velocity evolution versus the superficial gas velocity in the riser and the influence of the top clearance h_t on U_{Lr}. U_{Lr} increases with U_{gr} until $U_{gr} \approx 0.21$ m s^{-1}, and then reaches a plateau value for $U_{gr} > 0.21$ m s^{-1}. In our reactor configuration, gas bubbles were trapped in the downcomer and therefore, U_{Lr} reached a plateau. Couvert et al. [6] and Livingston and Zhang [32] observed the same phenomenon. However, in their cases, the plateau was obtained for $U_{gr} > 0.015$ m s^{-1} and $U_{gr} > 0.02$–0.03 m s^{-1}, respectively.

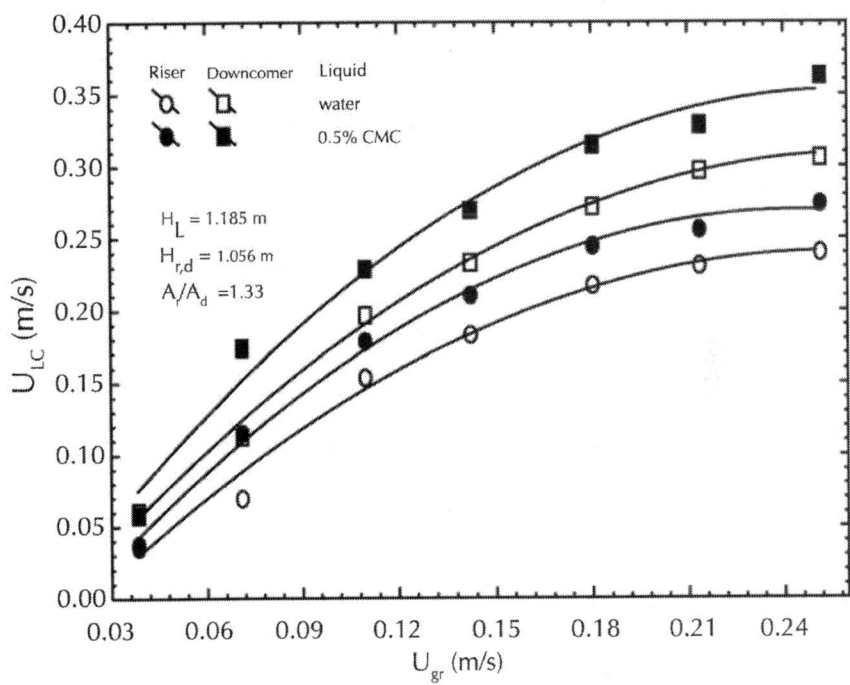

Figure 7: Influence of the gas superficial velocity on riser and downcomer liquid velocities, for $h_t = 0.265$ m and $h_b = 0.060$ m and two liquids (water, 0.15 M NaCl–0.5% CMC solution).

Fig. 8 shows a small but distinct influence of the baffle bottom clearance on the superficial riser liquid. The liquid velocity increases to a maximum with increasing h_b in the range $h_b < 0.076$ m (or $h_b/D_{hr} = 0.86$) and again diminishes with further increase of h_b. Kochbeck and Hempel [12] and Bando et al. [10], [11] and [33] observed the same trend. However, these results differed slightly with those reported in this work and by Koide et al. [13].

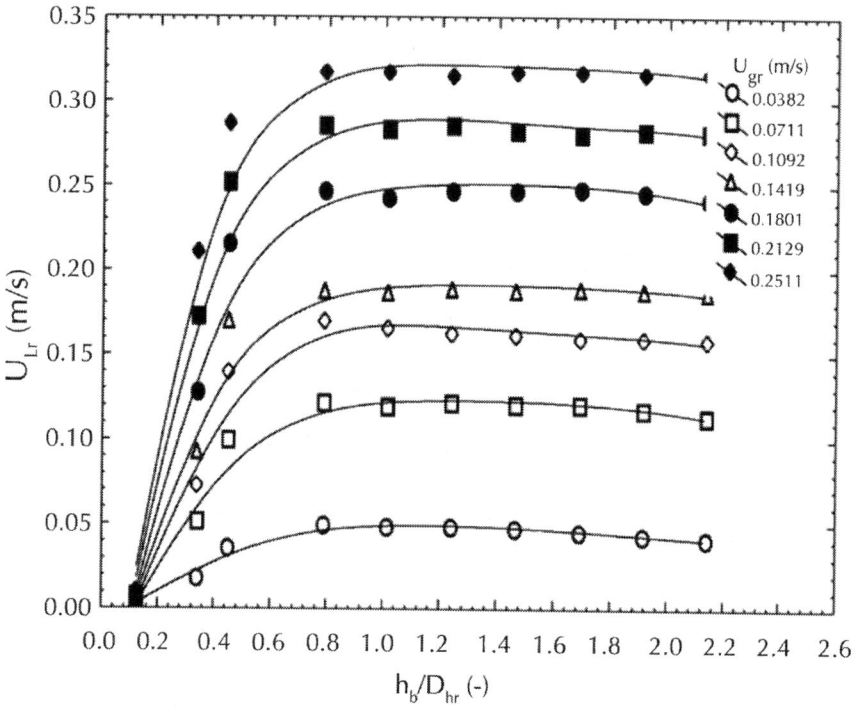

Figure 8: Plot of liquid circulations velocity against difference in gas holdups between the riser and the down comer for air–water system and h_b = 0.265 m at various gas superficial velocities and top clearances.

Fig. 9 shows the effect of h_t on U_{Lr} for the air–water system. The liquid circulation velocity increases with increasing h_t, and becomes unchanged when h_t is beyond a critical value $h_{t,crt}$ = 0.175 m for the column hydraulic diameter of D_{hc} = 0.10 m. The flow resistance to the reverse direction in the region above the upper end of the baffles is large at relatively short h_t and it decreases with increasing h_t. When h_t is beyond the critical value, the flow resistance becomes constant. From Fig. 9, it is considered that the critical value of h_t is a function of the column diameter and independent of the diameter and length of the riser, gas velocity and bubble size (d_B).

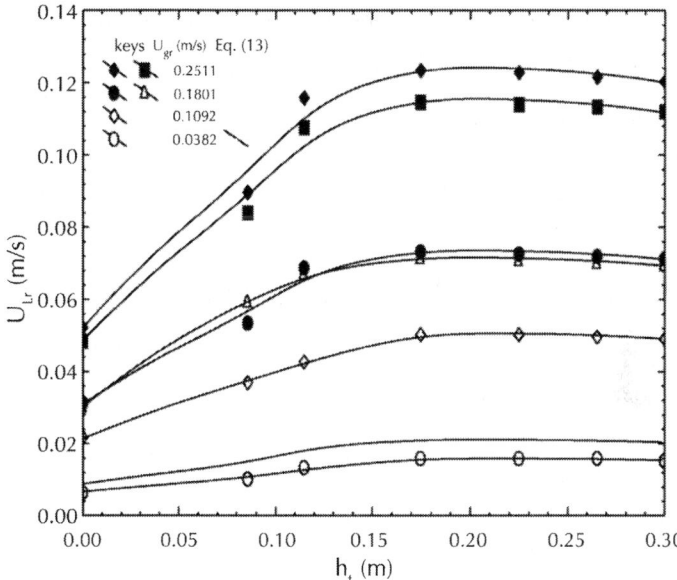

Figure 9: Effect of h_t on U_{Lr} for the air–water system and $h_b = 0.265$ m at various gas superficial velocities.

CORRELATION OF THE EXPERI-MENTAL RESULTS

Gas Holdup

Merchuk et al. [7] and Koide et al. [13] and [14] obtained semi-empirical correlations for gas holdups in an internal loop airlift reactor using water and CMC solutions. On the basis of their information, the following correlations were obtained from 56 data points to predict the gas holdup for each hydrodynamic region of the ALR proposed in this study.

Correlation ($R = 0.998$; $R^2 = 0.996$; S.D. = 0.075; S.T.D. error = 0.0078) for riser gas holdup E_r:

$$E_r = 770 \left(\frac{U_{gr}}{\sqrt{gD_{hc}}} \right)^{1.621} \left(\frac{g\eta_L^4}{\rho_L \sigma_L^3} \right)^{0.852} \left(\frac{h_b}{D_{hr}} \right)^{0.180}$$

$$\times \left(\frac{D_{es}}{D_{hc}} \right)^{11.375} \left(\frac{h_t}{D_{hr}} \right)^{5.2} \tag{8}$$

Correlation ($R = 0.952$; $R^2 = 0.888$; S.D. = 0.051; S.T.D. error = 0.0037) for downcomer gas holdup E_d:

$$E_d = 570 \left(\frac{U_{gr}}{\sqrt{gD_{hc}}} \right)^{2.82} \left(\frac{g\eta_L^4}{\rho_L \sigma_L^3} \right)^{0.968} \left(\frac{h_b}{D_{hr}} \right)^{0.210}$$

$$\times \left(\frac{D_{hs}}{D_{hc}} \right)^{11.375} \left(\frac{g\rho_L^2 D_{hc}^3}{\eta_L^2} \right)^{0.2} \tag{9}$$

Correlation ($R = 0.997$, $R^2 = 0.993$; S.D. = 0.092; S.T.D. error = 0.0075) for total gas holdup E_s:

$$E_s = 800.5 \left(\frac{U_{gr}}{\sqrt{gD_{hc}}} \right)^{1.73} \left(\frac{g\eta_L^4}{\rho_L \sigma_L^3} \right)^{0.852} \left(\frac{h_b}{D_{hr}} \right)^{0.08}$$

$$\times \left(\frac{D_{hs}}{D_{hc}} \right)^{11.5} \left(\frac{h_t}{D_{hr}} \right)^{4.82} \tag{10}$$

with the correlation of riser gas holdup, Eq. (8). Correlations (8), (9) and (10) which are based on data within the ranges:

$$\left. \begin{array}{l} 0.0008 < \dfrac{(U_{gr})}{(\sqrt{gD_{hc}})} < 0.45 \\[2mm] 2 \times 10^8 < \dfrac{(g\rho_L^2 D_{ec}^3)}{\eta_L^2} < 8 \times 10^{15} \\[2mm] 4 \times 10^5 < \dfrac{(g\eta_L^4)}{\rho_L \sigma_L^3} < 4 \times 10^{12} \\[2mm] 0.01 < \dfrac{h_b}{D_{hr}} < 0.1 \\[2mm] 0.1 < \dfrac{D_{hs}}{D_{hc}} < 0.5 \\[2mm] 0.045 < \sigma_L < 0.085 \, \text{N/m} \\[2mm] 0.02 < \eta_{eff} < 0.5 \, \text{Pas} \\[2mm] 0.33 \times 10^{-9} < D_L < 2.55 \times 10^{-9} \, \text{m}^2/\text{s} \end{array} \right\} \tag{11}$$

Eq. (10) is valid also for the total gas holdup E_T.

Fig. 10 compares the present and the reported data and reveals the roles played by the different regions in the rectangular airlift reactor. The relative contribution of the riser and downcomer to the separator and/or total gas holdups are different and complementary. The main term $((U_{gr})/(gD_{hc})^{0.5})$, which represents the influence of gas input rate and the liquid circulation rate, affects the holdup more strongly in the downcomer than in the riser. The behavior of the gas holdup in the gas separator is very close to that of the total gas holdup (Fig. 3). The bottom clearance, represented by the term (h_b/D_{hr}), exerts its influence on the gas holdup in all the regions. The exponent of $[h_b/D_{hr}]$ is positive in all the regions, indicating the effect of the liquid velocity. As h_b increases, the liquid circulation increases as well, and more gas bubbles are carried over into the downcomer. The exponent 0.21 on $[h_b/D_{hr}]$ in Eq. (9) indicates that E_d could be increased up to two-fold by changing the bottom clearance within the range of variables (U_{gr} = 0.0382 – 0.2511 m s^{-1}, h_b = 0.014 – 0.094 m) tested.

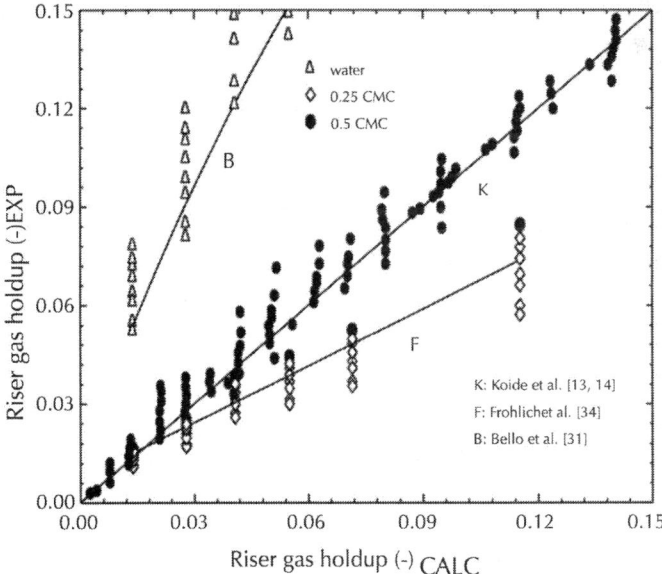

Figure 10: Parity plot of experimentally obtained values of the riser hold-

up vs. those calculated from Eq. (8) and comparison with data and correlations previously published.

The effect of the top clearance, represented by the group [h_t/D_{hr}], had a large influence on the riser and separator gas holdup, but none on the downcomer gas holdup. An increase in the top clearance decreases the gas holdup. The correlations show that the viscosity (η_L) and surface tension (σ_L), have a significant influence on gas holdups for the liquids tested and within the range of the variables inspected. They do significantly affect the downcomer gas holdup, because of the strong influence of free rising velocity of the bubbles ($U_{B\infty}$) on the bubble entrainment. A smaller influence is detected in the riser and separator gas holdup than in the downcomer. This is due to the fact that the downcomer holdup is smaller than in the riser and separator regions of the reactor.

The dimensional groups $\left[\left(g\rho_L^2 D_{hc}^3\right)/\left(\eta_L^2\right)\right]$ and $\left[\left(g\eta_L^4\right)/\left(\rho L\sigma_L^3\right)\right]$ do not show a statistically significant influence on the separator E_s gas holdup. At higher values of U_{gr} and the highest concentration of 0.15 M NaCl–CMC solution, considerably higher separator E_s gas holdup is observed (Fig. 5). This does not have much of an effect on $R^2 = 0.993$, but the correlation may underestimate the effect of viscosity and surface tension under the above conditions.

Fig. 10 shows data reported by Koide et al. [13] and [14]. These correlations fit those data satisfactorily. The data reported by Bello et al. [31] are also well represented by Eq. (10). In addition to these experimental data, Fig. 10 also shows the correlation by Frohlich et al. [34]. This correlation deviates slightly from the one presented here. It should also be noted that the Frohlich correlation requires the liquid velocity as an argument. In the present work, we took the velocity corresponding to the largest value of h_b, which is closer to the range of variables taken by Frohlich et al. [34].

Liquid Circulation Velocity

Lu et al. [16] and Bello et al. [31], correlated the liquid velocity as $U_{Lr} = \alpha \ (A_d/A_r)\beta \ (U_{gr})\delta(1 - E_r)$ with α, β, and $\delta = 0.124, 0.08, 0.537$

and 0.66, 0.78 ± 0.08, 0.33 for rectangular and concentric airlift reactors, respectively. Kockbeck and Hempel [12] correlated the liquid velocity as $U_{Lr} = (2gH_{r,d}E_r/[f_b(A_d/A_r)^2])^{1/2}$ with $f_b = 11.402 (A_d/A_b)y$ where $y = 0.789$. Fig. 11 shows that this correction cannot be applied for the investigated reactor configuration. In the present configuration, some bubbles are entrained and trapped in the downcomer and therefore, U_{Lr} reaches a plateau. The experimental and calculated values differ by up to ±29%.

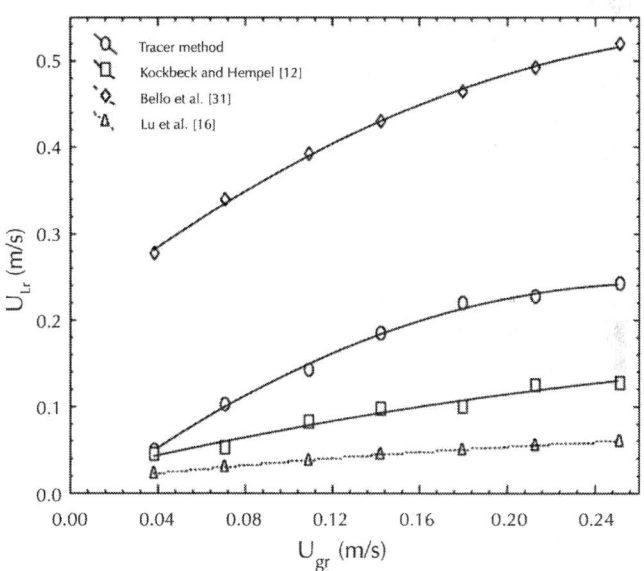

Figure 11: Bulk liquid velocity vs. U_{gr} for $h_t = 0.265$ m and $h_b = 0.060$ m and air–water system.

The pressure balance over the reactor gives an explanation for the above mentioned discrepancies. The liquid flow in the investigated rectangular airlift loop reactor was based on the pressure difference (P_{hydro}) between the riser and the downcomer.

$$\Delta P_{hydro} = \rho_{Lg}H_{r,d}(E_{gr} - E_{gd}) \tag{12}$$

The hydrostatic pressure is balanced by the dynamic pressure drop (ΔP_{dyn}, Eq. (13)), which is the sum of the pressure drops in the riser

(ΔP_r), downcomer (ΔP_d), bottom (ΔP_b), and top (ΔP_t) regions of the reactor (cf., Fig. 1).

$$\Delta P_{dyn} = \Delta P_r + \Delta P_d + \Delta P_b + \Delta P_t \qquad (13)$$

The details of the derivation of the final U_{Lr} (Eq. (14)) leading from Eqs. (12) and (13) are found elsewhere [12]. Eq. (14) is a modified version of the one presented by Kockbeck and Hempel [12], where f_t and f_b are the loss coefficients in the regions above the upper end of the baffles, respectively. Chisti et al. [35] have correlated f_b with the column bottom configuration.

$$U_{Lr} = \left[\frac{2gH_{r.d}(E_{gr} - E_{gd})}{f_r + f_t(2A_r/A_r + (A_r - A_d)[1 - (E_{gs} + E_{gr})])^2 + f_b(2A_r/A_d + (A_r - A_d)[1 - (E_{gr} + E_{gd})])^2} \right]^{0.5} \qquad (14)$$

$$f_b = 11.402 \left(\frac{A_d}{A_b} \right)^{0.789} \qquad (15)$$

Bando et al. [10] correlated f_t with riser, top clearance and column configurations using the following equations:

$$f_t = 10 \left(\frac{D_{hr}}{D_{hc}} \right)^{-5.6} \exp \left(-\frac{1.7h_t}{D_{hc}} \right), \quad \frac{h_t}{D_{hc}} < 2 \qquad (16)$$

and

$$f_t = 0.33 \left(\frac{D_{hr}}{D_{hc}} \right)^{-5.6}, \quad \frac{h_t}{D_{hc}} > 2 \qquad (17)$$

where A_d and A_b are cross-sectional areas of downcomer and bottom free zone between the riser and the downcomer. D_{hr} and D_{hc} are the equivalent hydraulic diameters of the riser and column, respectively. In the present study:

$$Ab = W_d h_b, \quad A_d = W_d L_d \qquad (18)$$

where h_b is considered constant and equal to 0.095 m, f_b equals to 43 and is constant also for all experimental conditions. From Eqs. (14), (15), (16) and (17), when f_t is constant, the liquid circulation velocity is regarded to be proportional to the square root of the gas holdup difference between the riser and the downcomer ($E_g = E_{gr} - E_{gd}$) and vice versa.

Fig. 12 and Fig. 13 shows the effect of bottom and top clearances, respectively, on liquid circulation velocity plotted against the gas holdup difference between riser and downcomer for the data observed in previous works [10], [12] and [35] and in this work. The U_{Lr} observed by Kockbeck and Hempel [12](h_b = 0.04 – 0.20 m, A_r/A_d = 6.4 – 17.4) roughly agree with those observed in this work. When the gas velocity is variable under constant h_t, the liquid circulation velocity is roughly linear to the square root of the gas holdup difference. On the other hand, when h_t and h_b are variable under the constant gas velocity, the data hardly ride on the lines with slope of 0.454 (Fig. 12) and 0.115 (Fig. 13), respectively. From this, the loss coefficients are considered to change due to the variable h_t and h_b.

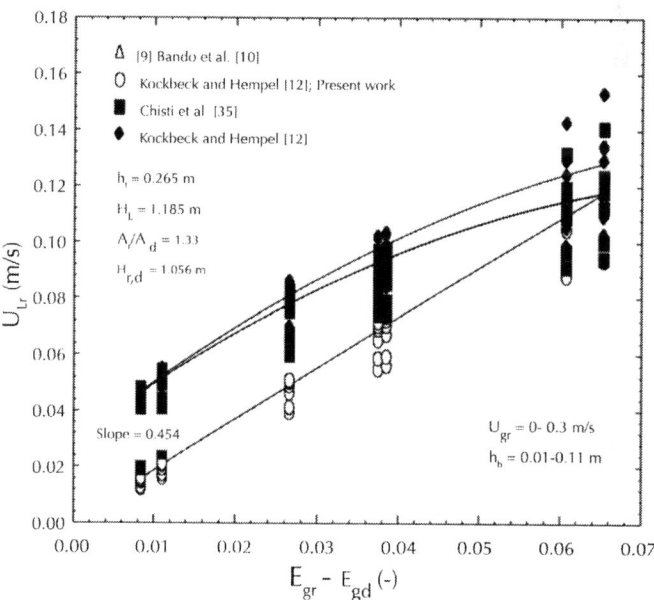

Figure 12: Effect of h_b on U_{Lr} for air–water system and h_t = 0.265 m at various gas superficial velocities.

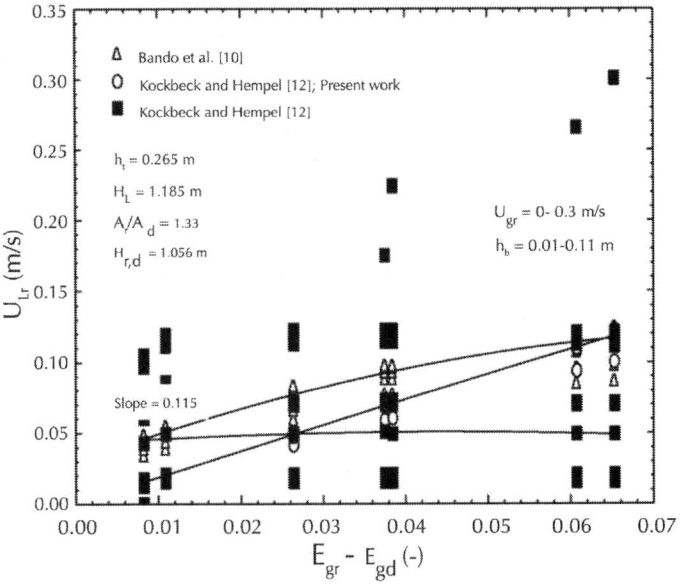

Figure 13: Effect of h_t on U_{Lr} for air–water system and h_b = 0.095 m at various gas superficial velocities.

CONCLUSIONS

Experiments have been carried out using water and mineralized CMC solutions of effective viscosity ranging from 0.02 to 0.5 Pa s, and surface tension from 0.065 to 0.085 N/m to investigate the influence of baffle (top and bottom) clearance design on gas holdup and liquid circulation velocity in a two riser rectangular airlift reactor with inverse internal loop and expanded gas–liquid separator. The gas holdups for each of the different hydrodynamic regions in the rectangular airlift reactor: riser, downcomer, and gas–liquid separator were successfully correlated using expressions derived through dimensional analysis for several bottom and top clearances.

The experimental results for the liquid circulation velocity were successfully correlated using empirical models obtained via pressure balance and loss coefficient. The calculated and measured values agreed within an error of ±29%. The liquid circulation velocity increased when the top and bottom clearances were increased and remained unchanged when h_b/D_{hr} and h_t were above 1.0 and 0.175 m, respectively. Both the bottom and top baffle clearances were found to influence liquid velocity to some degree. The bottom clearance was considered the most important characteristic.

The liquid circulation velocity data revealed that the design of the gas–liquid separator is a very important factor affecting the hydrodynamic performance. This has been already stated earlier by other investigators (Freitas and Teixeira [36], Siegel and Merchuk [37], and Vicent and Teixeira [38].

ACKNOWLEDGMENTS

We acknowledge the support by the Ontario Government Scholarship (OGS) awarded to Peter M. Kilonzo and the support by the Natural Science and Engineering Research Council of Canada (NSERC) awarded to Dr. A. Margaritis through individual Discovery Grants.

APPENDIX A. CALCULATION OF COEFFICIENTS

The experimentally measured data for the liquid velocities and gas holdups for the double riser rectangular airlift reactor with inverse internal-loop and expanded gas–liquid separator were correlated with the equation developed by [35]:

$$U_{Lr} = \left[\frac{2gH_{r,d}(E_{gr} - E_{gd})}{f_t'/(1 - E_{gr})^2 + f_b'(A_r/A_d)^2(1/(1 - E_{gd})^2)} \right]^{1/2} \tag{A1}$$

Note that Eq. (A1) contains both f_b and f_t because the connections of the riser and downcomer at the top and bottom sections were assumed to have different geometries. The values of fb and ft required to achieve best fit of the data with Eq. (A1) were calculated using the reactor geometry. The calculated f_b values were correlated with the geometric variables as using the equation:

$$P_{Tr} - P_{Td} = \frac{1}{2} \rho_L U_{Lr}^2 f_{tE} + \frac{1}{2} f_{tC} \rho_L U_{Ld}^2$$

(A2)

For the simulations reported in this work, the geometrical dependent frictional loss coefficient at the top section was evaluated equal to $f_{tC} = 14.06$. The calculated value was due to the fluid expansion from the riser region to the gas separation region f_{tE} and fluid contraction from the separator to the downcomer region f_{tC}. Their calculated values $f_{tE} = 0.82$ for the geometric variables (A_r/A_t) was correlated with the empirical Eq. (A3) [39] and [40]

$$f_{tE} = \left[1 - \left(\frac{A_r}{A_s} \right) \right]^2$$

(A3)

while $f_{tC} = 13.24$ was correlated with the geometric variables (A_t/A_d) [41] and [42] using the equation

$$f_{tC} = \left(\frac{A_s}{A_d} - 1 \right)$$

(A4)

where A_t is the cross-sectional area of the liquid level in the gas–liquid separator. The total top frictional coefficient was evaluated as a sum of the expansion and contraction terms at the top section as:

$f_t = f_t E + f_t C$

(A5)

A high value of 54.52 was used for the bottom loss coefficient f_b to simulate the restricted flow situation to account for the pressure drop through the gap between the baffle bottom and the base plate f_b and through the sparger system f_{bSP}. The value of fb=17.66 account for fluid contraction from the downcomer to the riser and was evaluated for the geometric ratio $A_d/A_b = 0.72$ having the bottom clearance $h_b = 0.06$ m. The calculated value was correlated with the empirical equation [5], [43] and [44]

$$f_b' = 11.402 \left(\frac{A_d}{A_b} \right)^{0.789}$$

(A6)

where A_b is the free cross-sectional area of fluid flow at the bottom of the baffle.

A value of $f_{bSP} = 36.86$ was obtained to account for the fluid contraction through the sparger tube system from the bottom of the riser having the geometric $A_r/A_{FSP} = 1.84$, and correlated with the equation:

$$f_{bSP} = 11.402 \left(\frac{A_r}{A_{FSP}} \right)^{0.789}$$

(A7)

where A_{FSP} is the free area of flow in the sparging regions, given by the equation:

$$A_{FSP} = A_r - A_{SP}$$

(A8)

The overall bottom coefficient f_b is the sum of the two coefficients accounting for the restricted flow through the baffle bottom f'b, and sparger tubes f_{bSP}, i.e.,

$$f_b = f_b' + f_{bSP}$$

(A9)

REFERENCES

1. Y. Chisti, Airlift Bioreactors, Elsevier, London/New York, 1989.

2. U. Onken, P. Weiland, Airlift fermenters: construction behaviour and uses, in: Adv. Biotechnol. Proc., vol. 1, Alan R. Liss, New York, 1983.

3. Y. Chisti, M. Moo-Young, Airlift bioreactorts for treatment of hydrocarbon contaminated wastes, in: W.K. Teo, M.G.S.

Yap, S.K.W. Oh (Eds.), Better Living through Biochemical Engineering, University of Singapore, Singapore, 1999.

4. P. Varey, Airlift for purity, Chemical Eng. (London) 529 (1992) 37.

5. W.-J. Lu, S.-J. Hwang, C.-M. Chang, Liquid velocity and gas holdup in three-phase internal loop airlift reactors, Chem. Eng. Sci. 30 (1995) 1301–1310.

6. A. Couvert, M. Roustan, P. Chatellier, Two-phase hydrodynamic study in a rectangular air-lift loop reactor with an internal baffle, Chem. Eng. Sci. 54 (1999) 5245–5252.

7. J.C. Merchuk, N. Ladwa, A. Cameron, M. Bulmer, A. Picket, Concentric-tube airlift reactors: Effects of geometrical design on performance, AIChE J. 40 (1994) 1105–1117.

8. V. Lazarova, J. Meyniel, L. Duval, J. Manem, A novel circulating bed reactor: hydrodynamics, mass transfer and nitrification capacity, Chem. Eng. Sci. 52 (1997) 3919–3927.

9. F. Yamashita, Gas holdup in bubble column with draft tube for gas dispersion annulus, Chem. Eng. Jpn. 31 (2) (1998) 289–294.

10. Y. Bando, K. Fujimori, H. Terazawa, K. Yasuda, M. Nakamura, Effects of equipment dimensions on circulation flow rates of liquid and gas in bubble column with draft tube, J. Chem. Eng. Jpn. 33 (3) (2000) 379–385.

11. Y. Bando, H. Hayakawa, M. Nakamura, Effects of equipment dimensions on liquid mixing time of bubble column with draft tube, J. Chem. Eng. Jpn. 31 (5) (1998) 765–770.

12. B. Kochbech, D.C. Hempel, Liquid velocity and dispersion coefficient in an airlift reactor with inverse internal loop, Chem. Eng. Technol. 17 (1994) 401–405.

13. K. Koide, M. Kimura, H. Nitta, H. Kawabata, Liquid circulation in bubble column with draught tube, J. Chem. Eng. Jpn. 21 (4) (1988) 393–399.

14. K. Koide, S. Iwamoto, T. Takasaka, S. Matsuura, E. Takahashi, M. Kimura, H. Kubota, Liquid Circulation, Gas holdup and

pressure drop in bubble column with draft tube, J. Chem. Eng. Jpn. 17 (6) (1984) 611–618.

15. K. Koide, H. Sato, H. Iwamoto, Gas holdup and volumetric liquid phase mass transfer coefficient in bubble column with draft tube and with gas dispersion into annulus, J. Chem. Eng. Jpn. 16 (1983) 407–413.

16. X. Lu, J. Ding, Y. Wang, J. Shi, Comparison of hydrodynamics and mass transfer characteristics, of a modified square airlift reactor with common airlift reactors, Chem. Eng. Sci. 55 (2000) 2257–2263.

17. P.-M. Wang, T.-K. Huang, H.-P. Cheng, Y.-H. Chien, W.-T. Wu, A modified airlift reactor with high capabilities of liquid mixing and mass transfer, J. Chem. Eng. Jpn. 35 (2002) 354–359.

18. P.M. Kilonzo, A. Margaritis, The effect of non-Newtonian fermentation broth viscosity and small bubble segregation on oxygen mass transfer in gas-lift bioreactors: a critical review, Biochem. Eng. J. 17 (2004) 27–40.

19. E.E. Petersen, A. Margaritis, Hydrodynamic and mass transfer characteristic of three-phase gaslift bioreactor systems, Crit. Rev. Biotechnol. 21 (4) (2001) 233–294.

20. M.K. Popovic, C.W. Robinson, Mass transfer studies of external-loop airlifts and bubble column, AIChE J. 35 (1989) 393–405.

21. P. Shamlou, D.J. Pollard, A.P. Ison, M.D. Lilly, Gas holdup and liquid circulation rate in concentric tube airlift bioreactors, Chem. Eng. Sci. 49 (3) (1994) 303–312.

22. M. Nishikawa, H. Kato, K. Hashimoto, Heat transfer in aerated tower filled with non-Newtonian liquids, Ind. Eng. Proc. Des. Dev. 16 (1977) 133–137.

23. M. Nakanoh, F. Yoshida, Gas absorption by Newtonian and nonNewtonian liquids in a bubble column, Ind. Eng. Chem. Process Des. Dev. 19 (1980) 190–195.

24. D.G. Allen, C.W. Robinson, Hydrodynamics and mass transfer in Aspergillus niger fermentations in bubble column and loop

bioreactors, Biotechnol. Bioeng. 34 (1989) 731–740.

25. G.G. Li, S.-Z. Yang, Z.-L. Cai, J.-Y. Chen, Mass transfer and gas–liquid circulation in an airlift bioreactor with viscous non-Newtonian fluids, Chem. Eng. J. 56 (1995) B101–B107.

26. M. Moo-Young, B. Hallard, D.G. Allen, R. Rurrel, Y. Kawase, Oxygen transfer to mycelial fermentation broths in an airlift fermenter, Biotechnol. Bioeng. 30 (1987) 746–753.

27. C. Vial, E. Camarasa, S. Poncin, G. Wild, N. Midoux, J. Bouillard, Study of hydrodynamic behavior in bubble columns and external loop airlift reactors through analysis of pressure fluctuations, Chem. Eng. Sci. 55 (2000) 2957–2973.

28. A.B. Russe, C.R. Thomas, M.D. Lilly, The influence of height and topsection size on the hydrodynamic characteristics of airlift fomenters, Biotechnol. Bioeng. 43 (1994) 69–76.

29. Y. Kawase, N. Omori, M. Tsujimura, Liquid–phase mixing in external-loop bioreactors, Chem. Technol. Biotechnol. 63 (1994) 49– 55.

30. Y. Wang, B. McNeil, A study of gas holdup, liquid velocity, and mixing time in a complex high viscosity, fermentation fluid in an airlift bioreactor, Chem. Eng. Technol. 19 (1996) 143–153.

31. R.A. Bello, C.W. Robinson, M. Moo-Young, Liquid circulation and mixing characteristics of airlift contactors, Can. J. Chem. Eng. 62 (1984) 573–577.

32. A.G. Livingston, S.F. Zhang, Hydrodynamic behavior of threephase (gas–liquid–solid) gaslift reactors, Chem. Eng. Sci. 48 (1993) 1641–1654.

33. Y. Bando, M. Nakamura, H. Sota, K. Toyoda, S. Sizuki, A. Idota, Flow characteristics of bubble column with perforated draft tube—effect of equipment dimensions and gas dispersion, J. Chem. Eng. Jpn. 25 (1) (1992) 49–54.

34. S. Frohlich, M. Lotz, T. Korte, A. Lubbert, K. Schugerl, M. Seekamp, Characterization of a pilot plant airlift tower loop bioreactor. I. Evaluation of the two-phase properties with model media, Biotechnol. Bioeng. 38 (1991) 43–45.

35. M.Y. Chisti, B. Hallard, M. Moo-Young, Liquid circulation in airlift reactors, Chem. Eng. Sci. 43 (3) (1988) 451–457.

36. C. Freitas, J.A. Teixeira, Hydrodynamic studies in an airlift reactor with enlarged degassing zone, Bioprocess. Eng. 18 (1998) 267– 279.

37. M. Siegel, J.C. Merchuk, Hydrodynamics in rectangular airlift reactors, scale-up and the influence of gas–liquid separator design, Can. J. Chem. Eng. 69 (1991) 465–473.

38. A.A. Vicent, J.A. Teixeira, Hydrodynamic performance of three-phase airlift bioreactors with enlarged degassing zone, Bioprocess. Eng. 14 (1995) 17–22.

39. M.A. Young, R.G. Carbonell, D.F. Ollis, Airlift Bioreactors: analysis of local two-phase hydrodynamics, AIChE J. 37 (3) (1991) 403–428.

40. N. de Nevers, Fluid Mechanics for Chemical Engineers, McGraw-Hill, Inc., New York, 1991.

41. P. Verlaan, J. Tramper, K. VanT Riet, A hydrodynamic model for an airlift-loop bioreator with external loop, Chem. Eng. J. 33 (1986) B43–B53.

42. L.-S. Fan, S.-J. Hwang, A. Matsuura, Hydrodynamic behavior of a draft tube gas–liquid–solid spouted bed, Chem. Eng. Sci. 39 (12) (1984) 1677–1688.

43. Y. Chisti, M. Moo-Young, Improve the performance of airlift reactors, Chem. Eng. Prog. 89 (1993) 38–45.

44. Y. Chisti, M. Moo-Young, Communication to the editor on the calculation of shear rate and apparent viscosity in airlift and bubble column bioreactors, Biotechnol. Bioeng. 34 (1989) 1391–1392.

Citations

CHAPTER 1

Fatemeh Kavousi, Yaghoub Behjat, Shahrokh Shahhosseini, Optimal design of drainage channel geometry parameters in vane demister liquid–gas separators, Chemical Engineering Research and Design, Volume 91, Issue 7, July 2013, Pages 1212-1222, ISSN 0263-8762, http://dx.doi.org/10.1016/j.cherd.2013.01.012.

CHAPTER 2

Héctor Constant-Machado, Jean-Pierre Leclerc, Eddie Avilan, Gustavo Landaeta, Nelkys Añorga, Oscar Capote, Flow modeling of a battery of industrial crude oil/gas separators using 113mln tracer experiments, Chemical Engineering and Processing: Process Intensification, Volume 44, Issue 7, July 2005, Pages 760-765, ISSN 0255-2701, http://dx.doi.org/10.1016/j.cep.2004.08.005.

CHAPTER 3

L. Du, J. Zh. Yao, W.G. Lin, Experimental study of particle flow in a gas–solid separator with baffles using PDPA, Chemical Engineering Journal, Volume 108, Issues 1–2, 1 April 2005, Pages 59-67, ISSN 1385-8947, http://dx.doi.org/10.1016/j.cej.2004.12.043.

CHAPTER 4

T. Keller, J. Muendges, A. Jantharasuk, C.A. Gónzalez-Rugerio, H. Moritz, P. Kreis, A. Górak, Experimental model validation for n-propyl propionate synthesis in a reactive distillation column coupled with a liquid–liquid phase separator, Chemical Engineering Science, Volume 66, Issue 20, 15 October 2011, Pages 4889-4900, ISSN 0009-2509, http://dx.doi.org/10.1016/j.ces.2011.06.056.

CHAPTER 5

Shi-ying Shi, Jing-yu Xu, Huan-qiang Sun, Jian Zhang, Dong-hui Li, Ying-xiang Wu, Experimental study of a vane-type pipe separator for oil–water separation, Chemical Engineering Research and Design, Volume 90, Issue 10, October 2012, Pages 1652-1659, ISSN 0263-8762, http://dx.doi.org/10.1016/j.cherd.2012.02.007.

CHAPTER 6

Rainier Hreiz, Caroline Gentric, Noël Midoux, Richard Lainé, Denis Fünfschilling, Hydrodynamics and velocity measurements in gas–liquid swirling flows in cylindrical cyclones, Chemical Engineering Research and Design, Volume 92, Issue 11, November 2014, Pages 2231-2246, ISSN 0263-8762, doi.org/10.1016/j.cherd.2014.02.029.

CHAPTER 7

M. Creutz, D. Mewes, A novel centrifugal gas–liquid separator for catching intermittent flows, International Journal of Multiphase Flow, Volume 24, Issue 7, November 1999, Pages 1057-1078, ISSN 0301-9322, http://dx.doi.org/10.1016/S0301-9322(98)00018-4.

CHAPTER 8

Peter M. Kilonzo, Argyrios Margaritis, M.A. Bergougnou, JunTang Yu, Ye Qin, Influence of the baffle clearance design on hydrodynamics of a two riser rectangular airlift reactor with inverse internal loop and expanded gas–liquid separator, Chemical Engineering Journal, Volume 121, Issue 1, 1 August 2006, Pages 17-26, ISSN 1385-8947, http://dx.doi.org/10.1016/j.cej.2006.05.003.

Index